1 MONTH OF
FREE
READING

at

www.ForgottenBooks.com

By purchasing this book you are eligible for one month membership to ForgottenBooks.com, giving you unlimited access to our entire collection of over 1,000,000 titles via our web site and mobile apps.

To claim your free month visit:

www.forgottenbooks.com/free920025

ISBN 978-0-266-99069-7
PIBN 10920025

MENTAL ARITHMETIC,

UPON THE

INDUCTIVE PLAN;

BEING AN

ADVANCED INTELLECTUAL COURSE,

DESIGNED FOR

SCHOOLS AND ACADEMIES.

By BENJAMIN GREENLEAF, A.M.,
AUTHOR OF "NATIONAL ARITHMETIC," ETC.

IMPROVED EDITION.

BOSTON:
PUBLISHED BY ROBERT S. DAVIS & CO.
NEW YORK: D. APPLETON & CO., AND MASON BROTHERS.
PHILADELPHIA: J. B. LIPPINCOTT & CO.
CHICAGO: WILLIAM B. KEEN.
1860.

GREENLEAF'S SERIES OF MATHEMATICS.

1. PRIMARY ARITHMETIC; OR, MENTAL ARITHMETIC, upon the Inductive Plan; designed for Primary Schools. Improved edition, 72 pp.

2. INTELLECTUAL ARITHMETIC, upon the Inductive Plan; designed for Common Schools and Academies. Improved edition. 154 pp.

3. COMMON SCHOOL ARITHMETIC; OR, INTRODUCTION TO THE NATIONAL ARITHMETIC. Improved stereotype edition, 324 pp.

4. HIGHER ARITHMETIC; OR, THE NATIONAL ARITHMETIC, for advanced scholars in Common Schools and Academies. New electrotype edition, with additions and improvements. 444 pp.

5. PRACTICAL TREATISE ON ALGEBRA, for Academies and High Schools, and for advanced Students in Common Schools. Fifteenth improved stereotype edition. 360 pp.

6. ELEMENTS OF GEOMETRY; with PRACTICAL APPLICATIONS TO MENSURATION. Designed for Academies and High Schools, and for Advanced Students in Common Schools. Electrotype Edition. 320 pp. 12mo. Recently published.

COMPLETE KEYS TO THE INTRODUCTION, AND NATIONAL ARITHMETIC, AND THE PRACTICAL TREATISE ON ALGEBRA, containing Solutions and Explanations, for Teachers only. In 3 volumes.

☞ A KEY TO THE INTELLECTUAL ARITHMETIC, containing Answers and Solutions of the more difficult Examples, will be furnished to *Teachers only*, on application to the Publishers.

ELECTROTYPED BY HOBART & ROBBINS, BOSTON.

PREFACE.

THE object of this book is to furnish a properly graded course of higher Mental Arithmetic. It has therefore been the constant aim of the author, in its preparation, to unfold inductively the science of numbers in such a series of progressive intellectual exercises, as should awaken latent thought, encourage originality, give activity to invention, and develop the powers of discriminating justly, reasoning exactly, and of applying readily results to practical purposes.

The advanced exercises in the fundamental processes of the science, given toward the end of the book, constitute a feature peculiar to this work. These will be found useful, it is believed, as an intellectual drill, and also exceedingly valuable for preparing the learner to dispense with written operations in business life, to a far greater extent than has heretofore been deemed practicable.

In the notes, aid is furnished the pupil more by hints and suggestions, than by full and formal solutions, which, if too numerous, might discourage sufficiently persevering effort, and the all-important habit of self-reliance. Should, however, additional assistance appear to be required, in any case, intelligent teachers will, doubtless, feel it to be their particular province to furnish, in their own manner, the necessary explanations and illustrations.

BRADFORD, MASS., *September*, 1857.

ARITHMETICAL SIGNS.

A SIGN is a symbol employed to indicate the relations of quantities, or operations to be performed upon them.

1. The sign of *equality*, two short horizontal lines, =, is read *equal*, or *equal to*, and denotes that the quantities between which it is placed are equal the one to the other ; as 12 inches = 1 foot.

2. The sign of *addition*, an erect cross, +, is read *plus, and,* or *added to*, and denotes that the quantities between which it is placed are to be added together ; as $4 + 6$ equals 10.

3. The sign of *subtraction*, a short horizontal line, —, is read *minus*, or *less*, and denotes that the quantity on the right of it is to be subtracted from that on the left ; as $8 - 6$ equals 2.

4. The sign of *multiplication*, an inclined cross, ×, is read *times*, or *multiplied by*, and denotes that the quantities between which it is placed are to be multiplied together ; as 5×4 equals 20.

5. The sign of *division*, a horizontal line between two dots, ÷, is read *divided by*, and denotes that the quantity on the left is to be divided by that on the right ; as $18 \div 2$ equals 9.

SUGGESTIONS TO TEACHERS.

THE book should not be used by the class during recitation. Each question should be repeated by the pupil after the teacher, and the required solution given promptly.

No form of solution should be allowed to pass, unless it is neatly expressed, and is *entirely accurate*.

Classes in an advanced course of written arithmetic, and in algebra, that have not had a suitable preliminary training in mental arithmetic, may be greatly benefited by going through the more difficult intellectual exercises of this book, in connection with those branches.

MENTAL ARITHMETIC.

LESSON I.

1. John had 1 peach, and his father gave him 1 more; how many peaches then had he?

2. Susan has 2 books, and Mary has 1 book; how many books have they both?

3. If you had 2 cherries, and I should give you 2 more, how many cherries would you then have?

4. Lucy found 2 pins, and Sarah found 3 pins; how many did both find?

5. If you should recite 2 lessons to-day, and 4 more to-morrow, how many would you recite in all?

6. A lemon cost 2 cents, and an orange cost 5 cents; how many cents did both cost?

7. Gave for a pencil 2 cents, and for some paper 6 cents; what was the cost of both?

8. On one bush there are 2 roses, and on another there are 7 roses; how many on both bushes?

9. 2 boys and 8 boys are how many boys?

10. A farmer sold a lamb for 2 dollars, and a calf for 9 dollars; how many dollars did he get for both?

11. Alfred caught 3 birds, and Jason caught 1 bird; how many birds did they both catch?

12. James has 3 marbles, and Charles has 2 marbles; how many marbles have they both?

13. A man sold a pig for 3 dollars, and a sheep for 3 dollars; how many dollars did he receive for both?

14. Mary has 3 books, and Margaret has 4 books; how many books have they both?

15. 3 hats and 5 hats are how many hats?

16. Edward gave 3 cents for a postage stamp, and 5 cents for a box of wafers; how much did both cost?

17. Eliza is 3 years old, and Laura is 6; what is the sum of their ages?

18. A farmer has 3 cows in one field, and 7 in another; how many has he in both?

19. In a class there are 3 girls and 8 boys; how many are there in the class?

20. A boy found under one apple-tree 3 apples, and under another 9 apples; how many did he find in all?

21. If you have 4 chestnuts in one hand, and 1 chestnut in the other, how many have you in both hands?

22. Susan had 4 merit marks, and obtained 2 more; how many then had she?

23. George found 4 eggs in one nest, and 3 eggs in another; how many did he find in both?

24. A man bought a cord of wood for 4 dollars, and half a ton of coal for 4 dollars; how much did both cost him?

25. A lady paid 4 cents for a skein of silk, and 5 cents for a spool of cotton; how much did she pay for both?

26. Ella gave 4 cents for candy, and 6 cents for nuts; how much did she give for both?

27. Lucy, having given to a beggar 4 cents, found she then had left 7 cents; how many cents had she at first?

28. Alfred bought a hook for 4 cents, and a line for 8 cents; how much did both cost?

29. A farmer sold 4 cows, and then had 9 left; how many cows had he at first?

LESSON II.

1. 1 and 1 are how many?
2. 2 and 1 are how many?
3. 2 and 2 are how many?
4. 2 and 3 are how many?
5. 2 and 4 are how many?
6. 2 and 5 are how many?
7. 2 and 6 are how many?
8. 2 and 7 are how many?
9. 2 and 8 are how many?
10. 2 and 9 are how many?
11. 3 and 3 are how many?
12. 3 and 4 are how many?
13. 3 and 5 are how many?
14. 3 and 6 are how many?
15. 3 and 7 are how many?
16. 3 and 8 are how many?
17. 3 and 9 are how many?
18. 4 and 4 are how many?
19. 4 and 5 are how many?
20. 4 and 6 are how many?
21. 4 and 7 are how many?
22. 4 and 8 are how many?
23. 4 and 9 are how many?
24. Abby found 5 pins, and Jane found 1 more; how many did they both find?
25. Ellen had 5 chickens, and her father gave her 2 more; how many did she then have?
26. Mary gave 5 cents for tape, and 3 cents for thread; how much did she give for both?
27. George bought 5 marbles, and had 4 given him; how many then had he?
28. John gave to one school-mate 5 nuts, and to another the same number; how many did he give to both?

29. Olive had 5 pins on her cushion, and stuck on it 6 more ; how many then had she ?

30. If you spend 5 cents, and have 7 cents left, how many had you at first?

31. 5 oranges and 8 oranges are how many oranges?

32. Joseph, having lost 5 cents, had only 9 cents left ; how many had he at first?

33. Gave 6 cents for paper, and 1 cent for a pen ; how much did both cost?

34. If you had 6 apples, and should have 2 more given you, how many would you then have?

35. How many slates are 6 slates and 3 slates?

36. Gave 6 cents for paper, and 4 cents for quills ; how many cents were paid for both?

37. If you should give 6 dollars for a vest, and 5 dollars for a pair of boots, how much would both cost?

38. A bookseller sold in one day 6 books, and in another day 6 more ; how many did he sell in all?

39. A farmer sold 6 sheep, and retained 7 ; how many had he at first?

40. If a clock cost 6 dollars, and a table 8 dollars, what would be the cost of both?

41. A farmer has 6 cows in one pasture, and 9 in another ; how many has he in both?

42. Paid 7 cents for a ruler, and 1 cent for a pencil ; what did both cost?

43. If a paper of pins cost 7 cents, and a pencil 2 cents, how many cents must be paid for both?

44. If 7 birds are upon a gate, and 3 upon the ground, how many are there in all?

45. 7 books and 4 books are how many books?

46. 7 horses and 5 horses are how many horses?

47. Laura had 7 needles, and her sister gave her 6 more ; how many then had she?

LESSON III.

1. 5 and 5 are how many?
2. 5 and 6 are how many?
3. 5 and 7 are how many?
4. 5 and 8 are how many?
5. 5 and 9 are how many?
6. 6 and 6 are how many?
7. 6 and 7 are how many?
8. 6 and 8 are how many?
9. 6 and 9 are how many?
10. 3 and 7 are how many?
11. 7 and 4 are how many?
12. 7 and 6 are how many?
13. 5 and 7 are how many?
14. 7 and 8 are how many?
15. 9 and 1 are how many?
16. 8 and 7 are how many?
17. William bought 7 marbles, and had given him 7 more; how many then had he?
18. If Charles has two notes due him, one for 7 dollars, and the other for 8 dollars, how much is due him in all?
19. There are on one side of a room 7 chairs, and on the other side 9; how many are there on both sides?
20. John had 8 cents, and his father gave him 1 more; how many cents then had he?
21. James has 8 sheep, and 2 lambs; how many has he of both?
22. Paid 8 dollars for sugar, and 3 dollars for salt; how much was paid for both?
23. In a certain class there are 8 boys, and 4 girls; how many are there in all?
24. Gave 8 dollars for wood, and 5 dollars for coal; how much did both cost?

25. If a coat cost 8 dollars, and a vest 6 dollars, how much will both cost?

26. If you had 8 dollars, and your father should give you 7 more, how many would you have?

27. If a barrel of flour cost 8 dollars, and half a barrel of beef cost 8 dollars, how much will both cost?

28. Bought a hundred weight of sugar for 8 dollars, and a quantity of butter for 9 dollars; what did the whole cost?

29. Levi earned in one week 9 dollars, and in another week only 1 dollar; how much did he earn in all?

30. Paid 9 dollars for a saddle, and 2 dollars for a bridle; how much did both cost?

31. Andrew paid 9 cents for a quart of nuts, and 3 cents for candy; how much did the whole cost him?

32. James raised 9 melons, and his brother raised 4; how many did they both raise?

33. There are 9 apples on one table, and 5 on another; how many are there on both tables?

34. If you should have 9 pens given you of one kind, and 6 of another kind, how many pens would you then have?

35. On one side of a pond are 9 lilies, and on the other 6; how many are there on both sides?

36. Edward walked 9 miles in one day, and 7 miles the next day; how many miles did he walk in all?

37. A farmer sold 9 pounds of butter at one time, and 8 pounds at another; how many pounds did he sell in all?

38. A miller ground 9 bushels of wheat, and 9 bushels of corn; how many bushels did he grind in all?

LESSON IV.

1. 2 and 8 are how many?
2. 2 and 6 are how many?
3. 6 and 8 are how many?
4. 3 and 5 are how many?
5. 5 and 5 are how many?
6. 4 and 3 are how many?
7. 3 and 9 are how many?
8. 9 and 2 are how many?
9. 7 and 7 are how many?
10. 1 and 3 are how many?
11. Three and six are how many?
12. Four and five are how many?
13. Five and one are how many?
14. Three and eight are how many?
15. Nine and three are how many?
16. Nine and four are how many?
17. Eight and two are how many?
18. Five and seven are how many?
19. One and eight are how many?
20. Two and two are how many?
21. Eight and six are how many?
22. Two and seven are how many?
23. Nine and five are how many?
24. Two and three are how many?
25. Three and four are how many?
26. Four and two are how many?
27. How many are four and seven?
28. How many are four and eight?
29. How many are four and nine?
30. How many are six and nine?
31. How many are six and six?
32. How many are eight and three?
33. How many are eight and eight?
34. How many are eight and seven?

35. How many are five and six ?
36. How many are five and eight ?
37. How many are nine and seven ?
38. How many are nine and six ?

LESSON V.

1. John has 3 marbles, Samuel has 5, and Jacob has 4 ; how many have they all ?

2. Gave 3 cents to Susan, 4 to Emily, and 2 to Ann ; how many cents were given to them all ?

3. Gave 4 nuts to one boy, 2 to another, and 4 to another ; how many nuts were given to the three boys ?

4. Bought a pound of sugar for 9 cents, a pound of raisins for 7 cents, and an ounce of nutmegs for 6 cents ; what was the cost of the whole ?

5. James is 4 years old, Edward 6, and Charles 8 ; what is the sum of their ages ?

6. Bought a sheep for 9 dollars, a lamb for 2 dollars, and a pig for 5 dollars ; what did the whole cost ?

7. Paid for wood 7 dollars, for coal 6 dollars, and for a saw 2 dollars ; how much did the whole cost ?

8. Charles had 3 peaches, and received 3 more from Albert, and 2 more from Edmund ; how many then had he ?

9. A farmer sold 6 bushels of wheat, 7 bushels of rye, and 8 bushels of corn ; how many bushels did he sell ?

10. A lady expended for silk 4 dollars, for gloves 1 dollar, and for a bonnet 9 dollars ; how many dollars did she expend in all ?

11. Bought a barrel of flour for 6 dollars, a barrel of apples for 3 dollars, and a keg of molasses for 8 dollars ; what was the cost of the whole ?

12. John spends 9 cents for paper, 3 cents for wafers, and 5 cents for pens; how many cents does he spend in all?

13. George buys 6 oranges at one time, 2 at another, and 8 at another; how many oranges did he buy in all?

14. 5 dollars, 9 dollars, and 4 dollars, are how many dollars?

15. How many are 1, 7 and 6? 3, 5 and 4? 3, 4 and 2?

16. Susan has 3 birds, Abby 4, and Ellen 9; how many have they all? 3, 4 and 9 are how many?

17. How many pigeons are 10 pigeons and 2 pigeons? 10 pigeons and 3 pigeons?

18. Gave 10 cents for a pine-apple, 6 cents for a ball, and 4 cents for a ruler; how many cents were given for the whole?

19. How many are 9, 7 and 6? 4, 8 and 6? 9, 2 and 5?

20. Expended for a coat 10 dollars, for a hat 5 dollars, and for a vest 3 dollars; how much was expended for the whole?

21. How many are 10 and 5? 10 and 6? 10 and 7? 10 and 9? 10 and 8?

22. If your brother should give you 10 apples, your uncle 4 apples, and your father 3 apples, how many apples would you then have?

23. Three boys, James, Henry, and Charles, went a fishing; James caught 10 fishes, Henry 8, and Charles 6; how many did they all catch? 10, 8 and 6 are how many?

24. Lucy bought some pins for 10 cents, some thread for 9 cents, and some tape for 4 cents; how much did they all cost? 10, 9 and 4 are how many?

LESSON VI.

1.   How many are 1 and 10?   1 and 20?   1 and 30?   1 and 40?   1 and 50?   1 and 60?   1 and 70?   1 and 80?   1 and 90?

2.   How many are 1 and 11?   1 and 22?   1 and 33?   1 and 44?   1 and 55?   1 and 66?   1 and 77?   1 and 88?   1 and 99?

3.   How many are 2 and 11?   2 and 21?   2 and 31?   2 and 41?   2 and 51?   2 and 61?   2 and 71?   2 and 81?   2 and 91?

4.   How many are 2 and 9?   2 and 19?   2 and 29?   2 and 39?   2 and 49?   2 and 59?   2 and 69?   2 and 79?   2 and 89?   2 and 99?

5.   How many are 3 and 8?   3 and 18?   3 and 28?   3 and 38?   3 and 48?   3 and 58?   3 and 68?   3 and 78?   3 and 88?   3 and 98?

6.   How many are 4 and 3?   4 and 13?   4 and 23?   4 and 33?   4 and 43?   4 and 53?   4 and 63?   4 and 73?   4 and 83?   4 and 93?

7.   How many are 5 and 5?   5 and 15?   5 and 25?   5 and 35?   5 and 45?   5 and 55?   5 and 65?   5 and 75?   5 and 85?   5 and 95?

8.   How many are 6 and 4?   6 and 14?   6 and 24?   6 and 34?   6 and 44?   6 and 54?   6 and 64?   6 and 74?   6 and 84?   6 and 94?

9.   How many are 7 and 6?   7 and 16?   7 and 26?   7 and 36?   7 and 46?   7 and 56?   7 and 66?   7 and 76?   7 and 86?   7 and 96?

10.   How many are 8 and 7?   8 and 17?   8 and 27?   8 and 37?   8 and 47?   8 and 57?   8 and 67?   8 and 77?   8 and 87?   8 and 97?

11.   How many are 9 and 4?   9 and 14?   9 and 24?   9 and 34?   9 and 44?   9 and 54?   9 and 64?   9 and 74?   9 and 84?   9 and 94?

12.   How many are 6 and 6?   6 and 16?   6 and

26 ?   6 and 36 ?   6 and 46 ?   6 and 56 ?   6 and
66 ?   6 and 76 ?   6 and 86 ?   6 and 96 ?

13.   How many are 3 and 5 ?   3 and 15 ?   3 and
25 ?   3 and 35 ?   3 and 45 ?   3 and 55 ?   3 and
65 ?   3 and 75 ?   3 and 85 ?   3 and 95 ?

14.   How many are 7 and 3 ?   7 and 13 ?   7 and
23 ?   7 and 33 ?   7 and 43 ?   7 and 53 ?   7 and
63 ?   7 and 73 ?   7 and 83 ?   7 and 93 ?

15.   How many are 2 and 6 ?   2 and 16 ?   2 and
26 ?   2 and 36 ?   2 and 46 ?   2 and 56 ?   2 and
66 ?   2 and 76 ?   2 and 86 ?   2 and 96 ?

16.   How many are 3 and 3 ?   3 and 13 ?   3 and
23 ?   3 and 33 ?   3 and 43 ?   3 and 53 ?   3 and
63 ?   3 and 73 ?   3 and 83 ?   3 and 93 ?

17.   How many are 8 and 8 ?   8 and 18 ?   8 and
28 ?   8 and 38 ?   8 and 48 ?   8 and 58 ?   8 and
68 ?   8 and 78 ?   8 and 88 ?   8 and 98 ?

18.   How many are 8 and 9 ?   8 and 19 ?   8 and
29 ?   8 and 39 ?   8 and 49 ?   8 and 59 ?   8 and
69 ?   8 and 79 ?   8 and 89 ?   8 and 99 ?

19.   How many are 9 and 10 ?   9 and 20 ?   9 and
30 ?   9 and 40 ?   9 and 50 ?   9 and 60 ?   9 and
70 ?   9 and 80 ?   9 and 90 ?

20.   How many are 10 and 10 ?   10 and 20 ?
10 and 30 ?   10 and 40 ?   10 and 50 ?   10 and 60 ?
10 and 70 ?   10 and 80 ?   10 and 90 ?

21.   How many are 9 and 9 ?   9 and 19 ?   9 and
29 ?   9 and 39 ?   9 and 49 ?   9 and 59 ?   9 and
69 ?   9 and 79 ?   9 and 89 ?   9 and 99 ?

22.   How many are 10 and 11 ?   10 and 21 ?
10 and 31 ?   10 and 41 ?   10 and 51 ?   10 and 61 ?
10 and 71 ?   10 and 81 ?   10 and 91 ?

23.   How many are 10 and 12 ?   10 and 22 ?
10 and 32 ?   10 and 42 ?   10 and 52 ?   10 and 62 ?
10 and 72 ?   10 and 82 ?   10 and 92 ?

24.   How many are 9 and 11 ?   9 and 31 ?   9
and 43 ?   9 and 86 ?   9 and 97 ?

LESSON VII.

1. Charles gave 15 cents for an arithmetic, 10 cents for a grammar, and 8 cents for a writing-book; what was the cost of the whole?

2. $15 + 10 + 8$ are how many?

3. Bought 12 bales of cotton, 6 bags of rice, and 2 boxes of sugar; what was the number of articles purchased?

4. $12 + 6 + 2$ are how many?

5. A lady purchased some silk for 25 dollars, and a shawl for 5 dollars; how much did she give for both?

6. A man bought a cow for 35 dollars, and a calf for 5 dollars; how much did both cost?

7. $35 + 5$ are how many?

8. If a wagon is worth 50 dollars, a saddle 9 dollars, and a whip 1 dollar, what is the value of the whole?

9. A boy found 62 chestnuts ~~under~~ one tree, and 20 under another; how many did he find in all?

10. $62 + 10 + 10$ are how many?

11. $34 + 26 + 6$ are how many?

12. A gentleman gave 46 dollars for a watch, 7 dollars for a chain, and 2 dollars for a key; how many dollars did he pay for the whole?

13. $46 + 7 + 2$ are how many?

14. Lucy had 70 pins in a cushion, and put in 20 more; how many had she then in the cushion?

15. I gave 80 apples to Peter, and had 20 apples left; how many did I have at first?

16. $60 + 20$ are how many?

17. $35 + 20 + 15$ are how many?

18. New York has 59 counties, Delaware has 3, and Rhode Island 5; how many counties have the three states?

19. A man gave 64 dollars for a piece of land; it cost him 10 dollars to fence it, and 2 dollars to have it ploughed; what was the whole cost?

20. A farmer raised 40 bushels of oats, 50 bushels of corn, and 20 bushels of turnips; how many bushels in all did he raise?

21. Rufus received 40 cents on his birthday, and 40 cents at Christmas; how many cents did he receive in all?

22. A farmer kept his sheep in 4 pens; in the first there were 20, in the second there were 10, in the third there were 8, and in the fourth there were 6; how many sheep did he have?

23. My book-case has 4 shelves; the first shelf contains 16 books, the second 12 books, the third 7 books, and the fourth 6 books; how many books are there in the book-case?

24. A man started on a journey; the first day he travelled 30 miles, the second day 10 miles, and the third day 9 miles; how many miles did he travel?

25. Edward bought a vest for 98 cents, some buttons for 12 cents, and some thread for 6 cents; what was the whole cost?

26. $83 + 17 + 3 + 7$ are how many?

LESSON VIII.

1. If you have 2 apples, and give one of them to your brother, how many apples will you have left?

2. James had 3 pencils, and gave 1 away; how many pencils then had James?

3. Lucy had 4 books, and gave 2 of them to Jane; how many books then had she?

4. If I have 5 peaches, and should eat 3 of them, how many should I have left?

5. Charles had 6 doves, but the cat caught 3 of them ; how many then had he ?

6. Rufus caught 7 fishes, and threw 4 of them back into the water ; how many had he left ?

7. Lydia had 8 nuts, but has eaten 4 of them ; how many has she left ?

8. I had 9 sheets of paper, but have given 5 of them to Charles ; how many have I left ?

9. William had 8 pears, but has given 6 of them to his teacher ; how many pears has he left ?

10. William caught 10 squirrels, but 6 of them were allowed to escape ; how many were retained ?

11. Thomas had 12 cents, but has spent 7 of them ; how many has he left ?

12. Gave 13 dollars for a barrel of flour, and 8 dollars for a tub of butter ; how much more did the flour cost than the butter ?

13. A farmer sold 16 sheep, and 8 lambs ; how many more sheep did he sell than lambs ?

14. James had 13 chickens, but 6 were taken by the hawks ; how many remained ?

15. George planted 19 trees, but only 9 lived ; how many died ?

16. Gave 15 cents for oranges, and 9 cents for lemons ; how much more did the oranges cost than the lemons ?

17. Paid 20 cents for nails, and 10 cents for brads ; how much did the one cost more than the other ?

18. Bought a clock for 10 dollars, and sold it for 15 dollars ; what was gained ?

19. John found 12 eggs, and Samuel 8 ; how many more did John find than Samuel ?

20. A man bargained to labor 17 days, but left at the end of 11 days ; how many more days had he agreed to labor ?

21. Gave 11 dollars to Emily, and 7 dollars to

Betsey; how many more dollars were given to Emily than to Betsey?

22. Paid 20 cents for a penknife, and 12 cents for an inkstand; how much more did the penknife cost than the inkstand?

23. Sarah is 17 years old, and Isabel is 9 years old; what is the difference of their ages?

24. A farmer raised 25 bushels of beans, and 11 bushels of peas; how many more bushels of beans did he raise than of peas?

25. Henry had 35 pins, and lost 15 of them; how many had he left?

26. Mary went a shopping with 40 cents in her purse, and when she returned she had only 10 cents remaining; how many cents had she spent?

27. I had 14 oranges, and sold 7 of them; how many had I left?

28. Thomas had 18 birds, and 9 of them flew away; how many birds remained?

29. Henry had 12 quarts of berries, and sold 6 of them; how many had he left?

30. Thomas recited 25 perfect lessons, and William only 8; how many more did Thomas recite than William?

31. Bought a cow for 27 dollars, and sold her for 24 dollars; how much did I lose by the bargain?

32. Sold a lot of wood for 20 dollars, and received in payment some cloth worth 7 dollars; how much was still due?

33. Bought a wagon for 28 dollars, and sold it for 40 dollars; how much was gained by the bargain?

34. 40 less 28 are how many?

35. A man paid 14 dollars to one man, and 4 to another; how much more did he pay to one than the other?

36. From a vessel containing 40 gallons, 15 gallons leaked out ; how many gallons still remained ?

37. A boy counted his chickens one night, and found he had 19 ; he counted them the next morning, and found he had but 14 ; how many were missing ?

38. A horse travelled 40 miles one day, and 27 the next day ; how many more miles did he travel the first than the second day ?

39. Bought a carriage for 60 dollars, and a harness for 35 dollars ; how much more did the carriage cost than the harness ?

40. A cistern, which holds 90 gallons, was full in the morning, but at night there were but 30 gallons left ; how many gallons had leaked out ?

LESSON IX.

1. 4 less 2 are how many ?
2. 4 less 3 are how many ?
3. 5 less 3 are how many ?
4. 6 less 4 are how many ?
5. 7 less 5 are how many ?
6. 7 less 4 are how many ?
7. 8 less 6 are how many ?
8. 9 less 5 are how many ?
9. 10 less 9 are how many ?
10. 11 less 8 are how many ?
11. 13 less 9 are how many ?
12. 12 less 7 are how many ?
13. 14 less 6 are how many ?
14. 15 less 3 are how many ?
15. 16 less 6 are how many ?
16. 19 less 9 are how many ?
17. 21 less 10 are how many ?
18. 25 less 10 are how many ?
19. 29 less 9 are how many ?

20. 24 less 8 are how many?
21. 30 less 10 are how many?
22. 42 less 10 are how many?
23. 24 less 10 are how many?
24. 30 less 15 are how many?
25. '20 less 5 are how many?
26. How many are 24 — 12?
27. How many are 30 — 8?
28. How many are 40 — 2?
29. How many are 40 — 7?
30. How many are 50 — 7?
31. How many are 60 — 10?
32. How many are 37 — 10?
33. How many are 49 — 19?
34. How many are 33 — 13?
35. How many are 23 — 10?
36. How many are 43 — 10?
37. How many are 44 — 14?
38. How many are 34 — 4?
39. How many are 34 — 14?
40. How many are 53 — 23?
41. How many, are 50 — 9?
42. How many are 60 — 20?
43. How many are 70 — 20?
44. How many are 90 — 30?
45. How many are 80 — 40?
46. How many are 90 — 40?
47. How many are 92 — 40?
48. How many are 70 — 30?
49. How many are 75 — 25?
50. How many are 77 — 23?
51. How many are 81 — 31?
52. How many are 85 — 45?
53. How many are 92 — 12?
54. How many are 96 — 16?
55. How many are 100 — 20?
56. How many are 100 — 25?

LESSON X.

1. A farmer sold 10 sheep to one man, 6 to another, and had 12 left; how many had he at first?

2. 10 + 6 + 12 are how many?

3. A man had 16 dollars; he paid 4 dollars to one man, and 6 dollars to another man; how many dollars were left?

4. 16 — 4 — 6 are how many?

5. Sold a chest of tea for 25 dollars, which was 7 dollars more than it cost; how much did it cost?

6. Paid 12 dollars for a barrel of flour, and 9 dollars for a hundred weight of sugar; what was the cost of both, and how much more was paid for the flour than for the sugar?

7. John had 29 apples; he gave 9 to his brother, 7 to his sister, and the rest to his mother; how many did he give to his mother?

8. 29 — 9 — 7 are how many?

9. James bought a slate for 15 cents, some pencils for 6 cents; if he had at first 30 cents, how much more has he to spend?

10. Isabel is 25 years younger than her mother, who is 45 years old; what is Isabel's age?

11. How much less than 46 is 25 + 10?

12. How much less than 58 is 31 + 19?

13. How much less than 45 is 59 — 19?

14. How much greater than 49 is 63 — 13?

15. How much greater than 63 is 87 — 22?

16. How much more than 40 + 13 is 60?

17. How much less than 75 is 35 + 25?

18. John found that if he had caught 17 fishes more, he should have caught 50 fishes; how many did he catch?

19. Bought one barrel of pork for 18. dollars,

and another for 16 dollars ; and sold the whole for 40 dollars ; how much was the gain ?

20. George spent 19 cents for candy, and 21 cents for fruit ; how much more would he have to spend to make 50 cents?

21. Thirty-two, and eight, and five, less ten, are how many ?

22. Twenty-eight, and three, and nine, less twelve, are how many ?

23. Forty-one, and six, and three, and nine, less eleven, are how many ?

24. Sixty-four, and four, and six, and one, less fifteen, are how many ?

25. Nineteen, and eleven, and seven, and thirteen, less twenty, are how many ?

26. Fifty-nine, and nine, and seven, and five, less thirty, are how many ?

27. Seventy-seven, and seven, and six, and four, and eight, less twelve, are how many ?

28. Eighty-six, and fourteen, and twenty, less five, are how many ?

29. Ninety-nine, and eleven, and ten, and five, less twenty-one, are how many ?

30. Two men bought a horse, the one paying 110 dollars, and the other 30 dollars less ; how many dollars did both pay ?

31. Henry, who is 19 years old, is 10 years older than James, who is 13 years younger than Arthur ; required the ages of James and Arthur.

32. Sarah, 4 years ago, was 15 years old, and Mary, 8 years from now, will be 27 years old ; what is the difference in their ages ?

33. James found under one tree 20 apples, under another 16, under another 8 ; and having given his brother 6, eaten 5, and thrown away some that were defective, he had 20 left ; how many did he throw away ?

34. If you should earn 80 cents one day and spend 20 cents, and the next day should earn 60 cents and spend 40 cents, how much would you have left of the two days' earnings?

35. A man went a "shopping" with 20 dollars; on the way one man paid him 30 dollars, and another man 15 dollars; he bought at one store goods to the amount of 17 dollars, and paid a bill of 13 dollars at another, and, on the way home, paid the tax collector 14 dollars; how much money did he have on reaching home?

LESSON XI.

1. What cost 2 apples, at 2 cents apiece?

NOTE. — Since one apple costs 2 cents, 2 apples will cost 2 times 2 cents, which are 4 cents.

2. What cost 2 lemons, at 3 cents apiece?

3. What cost 2 oranges, at 4 cents apiece?

4. What cost 2 oranges, at 5 cents apiece?

5. What cost 2 pounds of rice, at 6 cents a pound?

6. Bought 2 writing-books, at 7 cents apiece; what did they cost?

7. If one pound of sugar cost 8 cents, what will 2 pounds cost?

8. If one pound of veal cost 9 cents, what will 2 pounds cost?

9. At 10 cents each, what will 2 spelling-books cost?

10. What cost 2 quarts of berries, at 11 cents a quart?

11. What cost 2 pine-apples, at 12 cents apiece?

12. What cost 3 pears, at 4 cents apiece?

13. What cost 3 quarts of milk, at 5 cents a quart?

14. What cost 3 yards of braid, at 6 cents a yard?

15. If a horse will trot 7 miles in one hour, how far will he trot in 3 hours?

16. If one pound of raisins cost 8 cents, what will 3 pounds cost?

17. Bought 3 yards of cloth, at 9 cents a yard; how much did it cost?

18. At 3 dollars a pair, what will 10 pairs of boots cost?

19. At 3 dollars apiece, what will 11 hats cost?

20. What cost 3 loaves of bread, at 12 cents a loaf?

21. What cost 4 bushels of cranberries, at 3 dollars a bushel?

22. At 4 dollars a yard, what cost 4 yards of broadcloth?

23. At 4 cents a pound, what cost 6 pounds of rice?

24. What cost 5 vests, at 4 dollars apiece?

25. What cost 4 cords of wood, at 7 dollars a cord?

26. What cost 10 skeins of silk, at 4 cents a skein?

27. At 4 cents apiece, what cost 8 oranges? 9 oranges? 11 oranges? 12 oranges?

28. How many are 5 times 5? 5 times 6? 5 times 8? 5 times 9?

29. At 5 cents apiece, what cost 7 lead-pencils? 10 lead-pencils? 11 lead-pencils? 12 lead-pencils?

30. When 6 dollars are paid for a cord of wood, what must be paid for 5 cords? For 6 cords? For 7 cords?

31. How many are 6 times 8? 6 times 9? 6 times 10? 6 times 11? 6 times 12?

32. If a horse travel 7 miles in 1 hour, how far

will he travel in 7 hours? In 9 hours? In 11 hours?

33. At 7 cents a pound, what will 8 pounds of beef cost? 10 pounds? 6 pounds? 12 pounds?

34. When flour is 8 dollars a barrel, how much must be paid for 5 barrels? 9 barrels? 8 barrels? 10 barrels? 11 barrels? 12 barrels?

35. At 9 cents a pound, what cost 6 pounds of coffee? 9 pounds? 11 pounds? 10 pounds? 12 pounds?

36. What cost 10 yards of cambric, at 12 cents a yard?

37. If a quantity of provisions will supply 11 men 8 days, how long will it supply one man?

38. William sold 10 doves, at 10 cents apiece; how much did he receive for them?

39. What will 11 dozen of eggs cost, at 10 cents a dozen? At 11 cents? At 12 cents?

40. If 12 men can do a piece of work in 12 days, how long will it take one man?

LESSON XII.

1. Four times three are how many?
2. Three times two are how many?
3. Seven times three are how many?
4. Four times four are how many?
5. Five times two are how many?
6. Three times seven are how many?
7. Five times four are how many?
8. Two times six are how many?
9. Six times five are how many?
10. Five times three are how many?
11. Six times four are how many?
12. Five times six are how many?
13. Eight times six are how many?
14. Seven times five are how many?

15. Six times six are how many?
16. Seven times four are how many?
17. Six times three are how many?
18. Seven times seven are how many?
19. Eight times five are how many?
20. Eight times four are how many?
21. Nine times three are how many?
22. Two times ten are how many?
23. Three times eight are how many?
24. Four times nine are how many?
25. Five times seven are how many?
26. Six times two are how many?
27. Seven times eight are how many?
28. Ten times three are how many?
29. Ten times six are how many?
30. Nine times four are how many?
31. Nine times five are how many?
32. Seven times naught are how many?
33. Three times three are how many?
34. Eleven times two are how many?
35. Twelve times one are how many?
36. Ten times seven are how many?
37. Nine times six are how many?
38. Twelve times two are how many?
39. Eleven times four are how many?
40. Ten times eight are how many?
41. Twelve times five are how many?
42. Nine times eight are how many?
43. Eight times eight are how many?
44. Seven times six are how many?
45. Eleven times three are how many?
46. Eleven times five are how many?
47. Ten times nine are how many?
48. Eleven times six are how many?
49. Twelve times four are how many?
50. Eleven times seven are how many?
51. Eleven times eight are how many?

52. Twelve times six are how many?
53. Twelve times seven are how many?
54. Twelve times nine are how many?

LESSON XIII.

1. At 9 dollars apiece, what will 8 ploughs cost?

2. If 5 men can reap a field in 10 days, how long will it take one man to reap it?

3. James bought 3 oranges, at 5 cents each, and 6 lemons, at 2 cents each; how much did the whole cost?

4. Susan bought 8 yards of cotton cloth, at 9 cents a yard, and 4 skeins of thread, at 2 cents a skein; what was the cost of the whole?

5. A lady bought 8 pounds of sugar, at 11 cents a pound, and paid 8 dozen of eggs, at 10 cents a dozen, and the remainder in money; how much money did she pay?

6. What cost 7 quarts of cherries, at 6 cents a quart?

7. How much more will 7 quarts of currants, at 7 cents a quart, cost, than 8 quarts of berries, at 6 cents a quart?

8. How much more is 8 times 8 than 7 times 9?

9. How much more is 8 times 9 than 10 times 7?

10. If two men start from the same place, and travel in opposite directions, the one travelling at the rate of 3 miles an hour, and the other at the rate of 4 miles an hour, how far apart will they be at the end of 5 hours?

11. If one man can make 6 pairs of shoes in one day, how many pairs can 9 men make?

12. What cost 2 tons of hay, at 15 dollars a ton?

13. Bought 6 cords of wood, at 8 dollars a cord, and handed in payment 5 ten-dollar bills; how much change should be received back?

14. What cost 3 barrels of beef, at 16 dollars a barrel?

15. If one quire of paper cost 20 cents, what will 3 quires cost? 4 quires? 5 quires?

16. At 17 cents a yard, what will 2 yards of calico cost? 3 yards?

17. How many are 9×4? 10×8?

18. How much more is 10×10 than 11×9?

19. Two men start 50 miles apart, and travel towards each other, the one at the rate of 4 miles an hour, and the other at the rate of 3 miles; how far apart will they be at the end of 5 hours?

20. In a certain orchard there are 10 rows of trees, with 11 trees in each row; how many trees are there in the orchard?

21. A farmer sold 9 sheep, at 5 dollars apiece, and 5 lambs, at 3 dollars apiece; how much did he get for them all?

22. A tailor has a piece of broadcloth containing 33 yards; if he should cut from it 13 yards, what will the remainder be worth, at 4 dollars a yard?

23. Bought 6 writing-books, at 8 cents apiece, and 5 more, at 7 cents apiece, and sold the whole for 93 cents; how much was made by the sale?

24. A farmer has his wheat in 5 bins, containing 10 bushels each; how much is the whole worth, at 2 dollars a bushel?

25. Five times twelve, less ten, plus fifteen, are how many?

26. Four times fifteen, less thirteen, plus six, plus twelve, are how many?

27. Five times twenty, less twenty-five, are how many?

28. Eight times eleven, plus twelve, less thirty, are how many?

29. Ten times eight, plus eighteen, plus twelve, less twenty, less six, are how many?

30. Four times sixteen, plus six, plus ten, less four times five, are how many?

31. Five times twenty-five, less fifty, plus seventeen, plus eight, are how many?

32. Eight times twenty-five, less five times ten, are how many?

33. If a man earns 50 dollars in 5 weeks, and pays of his earnings 3 dollars a week for board, how much will he have left?

34. If one plough is worth 3 cords of wood, how many cords will 15 ploughs cost?

35. If a man earns 100 cents a day, and pays out for family expenses 60 cents, how much will he have left at the end of 5 days?

36. By putting in the savings-bank 14 dollars a month, how much may be saved in 6 months?

37. For how much must I sell 6 cows, which cost 25 dollars each, to gain 25 dollars?

38. If a frog should be 12 days in getting out of a well, by leaping up 12 feet every morning, and falling back 4 feet every evening, how deep is the well?

39. How many are 5 times 16, plus 20? 4 times 25, minus 30?

40. How many are $20 \times 6 + 15$? $31 \times 6 - 11$? $27 \times 4 - 16$?

41. George has 15 marbles, and Lewis has 3 times as many, less 10; how many has Lewis?

42. Laura gathered 4 quarts of strawberries, and Mary gathered 3 times as many, less 2 quarts; how many did Mary gather?

43. What cost 28 pounds of flour, at 5 cents a pound?

44. What cost 4 pecks of potatoes, at 19 cents a peck ?

45. A person poured into a can 16 quarts of oil at 4 different times, and from the same filled 8 jugs, holding 8 quarts each, and 2 jugs, holding 10 quarts each ; how much oil remained in the can ?

46. If John has 3 chickens, Joseph 5 times as many, and Henry as many as John and Joseph both, plus 4, how many has William, who has 2 times as many as they all ?

47. Three boys had given them 75 nuts, Robert receiving 20, William 2 times as many as Robert lacking 15, and Ezra the remainder ; how many did Ezra receive ?

48. Alfred and Edward sold the same quantity of berries ; Alfred received 4 dimes and 4 three-cent pieces ; Edward received a fifty-cent piece and 1 dime, and paid back in making change 1 five-cent piece and 1 three-cent piece ; how much more money did the one obtain than the other ?

LESSON XIV.

1. At 2 cents each, how many peaches can be bought for 4 cents ?

NOTE. — Since one peach can be bought for 2 cents, as many peaches can be bought for 4 cents as 2 is contained times in 4, which are 2.

2. Harry had 6 chestnuts, which he gave in equal numbers to 2 of his brothers ; how many did each receive ?

3. Lucy divided 9 apples equally among her 3 sisters ; how many did she give to each of them ?

4. Thomas has 8 apples, which he wishes to give to 2 boys ; how many can he give to each of them ?

5. If you wish to give 8 oranges to 4 persons, how many can you give to each?

6. Mary has 10 pins in 2 cushions; if equally divided, how many are there in each cushion?

7. Thomas distributed 10 flowers equally among 5 of his playmates; how many did each receive?

8. If 12 dollars be divided equally among 6 men, how many will each receive?

9. James has 12 peaches, which he wishes to give to 4 of his companions; how many can he give to each?

10. At 5 shillings a bushel, how many bushels of apples can be bought for 15 shillings?

11. At 2 dollars a barrel, how many barrels of apples can be bought for 14 dollars?

12. When 20 dollars are given for 4 cords of wood, how much is paid for one cord?

13. How many skeins of silk, at 3 cents a skein, can be bought for 21 cents?

14. How many yards of broadcloth, at 5 dollars a yard, can be bought for 25 dollars?

15. At 6 cents a paper, how many papers of needles can be purchased for 24 cents?

16. At 9 dollars a barrel, how many barrels of flour can be bought for 18 dollars?

17. At 7 cents a pound, how many pounds of rice can be bought for 28 cents?

18. When 5 yards of cloth are bought for 20 dollars, what is the price a yard?

19. If a horse trot 6 miles an hour, how long will it take him to trot 30 miles?

20. How many pears, at 4 cents apiece, can you purchase for 16 cents?

21. If a man pays 35 cents for 5 pounds of nails, how much are the nails a pound?

22. At 6 dimes a bushel, how many bushels of corn can be bought for 36 dimes?

23. How many boxes of strawberries, at 3 dimes apiece, can be bought for 33 dimes?

24. How many yards of cloth, at 4 dollars a yard, can you buy for 40 dollars?

25. At 9 cents a quart, how many quarts of molasses can you purchase for 45 cents?

26. At 8 cents a paper, how many papers of pins may be bought for 32 cents?

27. Mary divided 42 apples equally between 7 companions; how many did she give to each?

28. In an orchard there are 56 trees, and there are 8 rows; how many trees are there in each row?

29. If it take 11 yards to make one dress, how many dresses can be made from 44 yards?

30. When coal is 10 dollars a ton, how many tons can be bought for 60 dollars?

31. At 9 cents a pound, how many pounds of veal may be bought for 72 cents?

32. At 7 shillings a day, how many days will it take a man to earn 49 shillings?

33. For 63 dollars how many ploughs can be bought, at 9 dollars each?

34. At 8 dimes a day, how long will a man be in earning 64 dimes?

35. At 5 cents a yard, how many yards of ribbon can be bought for 60 cents?

36. At 12 dollars a week, how long will it require to earn 72 dollars?

37. For 40 apples how many melons can be purchased, at the rate of 8 apples for one melon?

38. At 9 dimes each, how many turkeys can be purchased for 108 dimes?

39. If a train of cars move at the rate of 12 miles an hour, how many hours will it require to move 96 miles?

40. 3 heifers were bought for 48 dollars; how much were they apiece?

41. For 51 dollars how many acres of land can be bought, at the rate of 17 dollars an acre?

42. A school, consisting of 120 pupils, is divided into 10 equal classes; how many pupils are there in each class?

LESSON XV.

1. Nine are how many times 3?
2. Eleven are how many times 4?

NOTE. 11 are as many times 4 as 4 is contained in 11, which are 2 times and 3 over, or 2 times and 3 remainder.

3. Twelve are how many times 2? 3? 4?
4. Thirteen are how many times 2? 4? 5? 6? 7?
5. Fourteen are how many times 2? 7?
6. Fifteen are how many times 3? 5? 6?
7. Sixteen are how many times 2? 4? 8?
8. Seventeen are how many times 2? 3? 5? 8? 9?
9. Eighteen are how many times 2? 3? 6? 9? 8?
10. Nineteen are how many times 9? 8? 7? 6? 5?
11. Twenty are how many times 2? 4? 5? 10? 11?
12. Twenty-one are how many times 3? 7? 10? 12?
13. Twenty-two are how many times 2? 11?
14. Twenty-three are how many times 2? 10? 11? 9?
15. Twenty-four are how many times 4? 6? 8? 12?
16. Twenty-five are how many times 5? 6? 10? 12?
17. Twenty-six are how many times 2? 13?

18. Twenty-eight are how many times 2? 4? 7? 14?

19. Thirty are how many times 2? 3? 5? 6? 10? 15?

20. Thirty-one are how many times 3? 6? 9? 10?

21. Thirty-two are how many times 4? 8? 16?

22. Thirty-four are how many times 6? 8? 10? 17?

23. Thirty-five are how many times 5? 7? 8? 10?

24. Thirty-six are how many times 4? 6? 9? 12? 18?

25. Thirty-seven are how many times 8? 9? 15? 16?

26. Forty are how many times 4? 10? 15? 20?

27. Forty-two are how many times 6? 7? 10? 21?

28. Forty-five are how many times 5? 9? 10? 11? 15?

29. Forty-eight are how many times 6? 8? 12? 16?

30. Fifty are how many times 5? 7? 9? 10?

31. Fifty-one are how many times 3? 5? 17?

32. Fifty-four are how many times 6? 9? 10? 12? 18?

33. Fifty-six are how many times 6? 7? 8? 28?

34. Sixty are how many times 3? 4? 12? 15? 30?

35. Sixty-four are how many times 4? 8? 16? 32?

36. Sixty-five are how many times 5? 10? 12? 13?

37. Sixty-nine are how many times 6? 7? 20?

38. Seventy are how many times 2? 5? 10?

39. Seventy-two are how many times 2? 3? 8? 9? 12?

40. Seventy-four are how many times 2? 10? 12? 20?

41. Seventy-five are how many times 3? 5? 15? 25?

42. Seventy-seven are how many times 7? 9? 10? 11?

43. Eighty are how many times 4? 5? 8? 10? 20? 40?

44. Eighty-four are how many times 2? 4? 7? 12? 42?

45. Eighty-five are how many times 5? 8? 12? 27? 40?

46. Eighty-eight are how many times 4? 8? 10? 11? 20?

47. Ninety are how many times 2? 3? 5? 9? 30?

48. Ninety-two are how many times 2? 10? 15? 18? 46?

49. Ninety-five are how many times 5? 10? 11? 12?

50. Ninety-six are how many times 6? 8? 12? 48?

51. Ninety-nine are how many times 3? 9? 10? 33?

52. One hundred are how many times 4? 5? 20? 25? 50?

53. One hundred and eight are how many times 2? 9? 12? 25? 54?

54. One hundred and ten are how many times 5? 10? 11? 20? 55?

55. One hundred and twelve are how many times 2? 4? 25? 50?

56. One hundred and twenty are how many times 2? 5? 6? 8? 12? 30? 60?

57. One hundred and twenty-five are how many times 5 ? 12 ? 20 ? 25 ? 50 ?

58. One hundred and thirty are how many times 2 ? 5 ? 10 ? 13 ? 25 ? 65 ?

59. One hundred and thirty-two are how many times 2 ? 4 ? 6 ? 11 ? 12 ? 33 ? 66 ?

60. One hundred and forty-four are how many times 4 ? 6 ? 9 ? 12 ? 36 ? 72 ?

61. One hundred and fifty are how many times 3 ? 10 ? 15 ? 25 ? 30 ? 75 ?

LESSON XVI.

1. For 24 cents how many oranges can be bought, at 3 cents apiece ? At 4 cents apiece ? At 6 cents apiece ?

2. For 36 cents how many pounds of sugar can be bought, at 9 cents a pound ? At 12 cents a pound ?

3. If you have 30 cents, how many pencils can you buy, at 5 cents apiece ? At 6 cents apiece ?

4. How many tons of hay can be bought for 60 dollars, at 10 dollars a ton ? At 12 dollars a ton ? At 15 dollars a ton ?

5. If two boats are 50 miles apart, and the one gains on the other 5 miles an hour, in how many hours will they be together ?

6. If a man earns 8 dollars a week, how long will it take him to earn 48 dollars ? 64 dollars ?

7. How many cloaks, containing 9 yards, can be made from 63 yards of cloth ? From 72 yards ?

8. If 51 dollars be divided between 3 men, how many dollars will each receive ?

9. For 72 cents how many pounds of beef can be bought, at 8 cents a pound ? At 9 cents ? At 12 cents ?

10. If you should have 31 cents, how many

writing-books could you buy, at 8 cents each, and how many cents would you have left?

11. If you had 57 dollars, how many sheep could you buy, at 4 dollars each, and how many dollars would you have left?

12. For 84 dollars how many coats can be bought, at 7 dollars each? At 12 dollars each?

13. How many cords of wood can be bought for 56 dollars, at 7 dollars a cord? At 8 dollars a cord?

14. How many cords of wood can be bought for 100 dollars, at 4 dollars a cord? At 5 dollars a cord?

15. When a man, having 75 dollars, can buy 9 pigs, and have 3 dollars left, what is the cost of the pigs each?

16. At 10 dollars each, how many ploughs can be bought for 60 dollars?

17. When sugar is 9 cents a pound, how many pounds can be bought for 54 cents? For 63 cents? For 90 cents?

18. How many times $5 + 2$ in 21? In 35?

19. How many times $6 + 4$ in 50? In 70?

20. How many times $9 + 2$ in 66? In 77?

21. How many times 8 in $55 + 9$? In $63 + 9$?

22. When calico is 11 cents a yard, how many yards can be bought for 44 cents? 66 cents? 110 cents?

23. Jason had 52 apples, and found 8 more; he then divided the whole equally among 4 schoolmates; how many did he give to each?

24. A man had 25 cows, and bought 35 more; if he should put them all into 5 pastures, how many would there be in each pasture?

25. How many times 10 in $115 - 5$? In $107 - 7$?

26. How many times 3 in $69 - 3$? In $78 - 6$?

27. How many times 9 less 4 in 63 — 8 ?

28. How many times 18 less 12 in 90 — 30 ?

29. How many times 26 — 6 in 110 — 30 ?

30. How many times 29 — 4 in 137 — 12 ?

31. A farmer had 47 bushels of apples ; saving 12 bushels for his own use, he sold the rest in equal quantities to 5 persons ; how many bushels did he sell to each person ?

32. A boy on the way to market with 19 fishes, lost 4 of them ; the rest he sold for 75 cents ; how much apiece did he get for those he sold ?

33. William had 16 chickens, but a cat caught 4 of them ; the rest he sold for 60 dimes ; how much apiece did he get for those he sold ?

34. Thomas sold some nuts for 12 cents, some apples for 18 cents, and some peaches for 20 cents, and with the money bought writing-books at 10 cents apiece ; how many writing-books did he buy ?

35. A gentleman divided 120 dollars equally among his 3 sons and 2 daughters ; how many dollars did he give to each ?

36. George had 59 pears ; he gave 4 of them to one companion, 3 to another, 2 to another, and divided the remainder equally among his 10 classmates ; how many did each receive ?

37. At 15 cents a pound, how many pounds of beef can be bought for 45 cents ? For 90 cents ? For 150 cents ?

38. At 18 cents a dozen, how many dozen of eggs can be bought for 36 cents ? For 72 cents ? For 180 cents ?

39. When 9 shawls can be bought for 108 dollars, what are they apiece ?

40. James bought 4 dozen of lead-pencils, at 25 cents a dozen, and paid for them in apples, at 10 cents a dozen ; how many dozen of apples did the pencils cost ?

LESSON XVII.

1. How many lemons, at 4 cents apiece, will pay for 4 oranges, at 2 cents apiece?

2. How many pears, at 3 cents apiece, will pay for 2 melons, at 6 cents apiece?

3. 3 times 8 are how many times 6? 4?

4. 3 times 10 are how many times 6? 5?

5. At 2 dollars a bushel, how many bushels of wheat must be given for 4 barrels of flour, at 8 dollars a barrel?

6. 4 times 7 are how many times 2? 14?

7. 4 times 9 are how many times 3? 12?

8. 4 times 12 are how many times 3? 8?

9. 5 times 8 are how many times 4? 10?

10. 5 times 10 are how many times 2? 25?

11. How many yards of broadcloth, at 4 dollars a yard, should be received in payment for 10 sheep, at 6 dollars each?

12. At 8 cents a pound, how many pounds of sugar can be bought for 6 dozen of eggs, at 12 cents a dozen?

13. If you should sell 7 quarts of chestnuts, at 8 cents a quart, how many slates could you buy, at 14 cents apiece?

14. If you should sell 8 quarts of milk, at 5 cents a quart, how many yards of cotton cloth, at 10 cents a yard, could you take in pay?

15. 6 times 6 are how many times 3? 18?

16. 6 times 9 are how many times 2? 27?

17. 6 times 15 are how many times 3 times 10?

18. 7 times 10 are how many times 5 times 7?

19. 8 times 8 are how many times 4 times 4?

20. Bought 9 quarts of cranberries, at 10 cents a quart; to pay for them gave raisins worth 15 cents a pound; how many pounds did it take?

21. Bought 9 yards of broadcloth, at 8 dollars a yard, and paid for it with flour, at 12 dollars a barrel; how many barrels did it take?

22. 9 times 8 are how many times $24 \div 2$?

23. 9 times 11 are how many times $33 \div 3$?

24. 10 times 6 are how many times $20 \div 4$?

25. 11 times 10 are how many times $45 \div 9$?

26. 11 times 12 are how many times $42 \div 7$?

27. 12 times 7 are how many times $24 \div 6$?

28. 12 times 8 are how many times $54 \div 9$?

29. When oranges are 5 cents apiece, and pine-apples 10 cents apiece, how many oranges will cost as much as 6 pine-apples?

30. Bought 12 buffalo robes, at 12 dollars apiece, and paid for them with wood, at 6 dollars a cord; how many cords did it take?

31. How many yards of broadcloth, at 5 dollars a yard, will it take to pay for 11 tons of hay, at 10 dollars a ton?

32. A man bought 12 bushels of wheat, at 2 dollars a bushel, and 4 yards of cloth, at 3 dollars a yard, and paid for them in work, at 9 dollars a week; how many weeks did he work?

33. 2 men bought a cow for 40 dollars, and a horse for 80 dollars; they sold the cow for 30 dollars, and the horse for 100 dollars; what was each man's share of the gain?

34. How much butter, at 20 cents a pound, must be given for 8 yards of calico, at 15 cents a yard?

35. Bought 6 barrels of pork for 120 dollars; at how much a barrel must it be sold to gain 2 dollars a barrel?

36. Bought 15 yards of cloth, at 6 dollars a yard, and 4 yards more, at 5 dollars a yard, and paid for the whole in hay, at 11 dollars a load; how many loads did it take?

4

LESSON XVIII.

1. If 4 barrels of apples cost 8 dollars, what cost 5 barrels?

Note. — If 4 barrels cost 8 dollars, 1 barrel will cost one fourth * of 8 dollars, which is 2 dollars, and 5 barrels will cost 5 times 2 dollars, which are 10 dollars.

2. If 3 pounds of butter cost 36 cents, what cost 5 pounds?

3. When 10 cents are paid for 5 rolls of candy, how much must be paid for 6 rolls?

4. When 5 yards of broadcloth cost 15 dollars, what will 9 yards cost?

5. If 4 tons of hay cost 40 dollars, what cost 3 tons?

6. When 7 quarts of fruit bring 35 cents, how much do 5 quarts bring?

7. If 6 barrels of soap cost 24 dollars, what cost 8 barrels?

8. If 8 bushels of wheat cost 16 dollars, how much will 7 bushels cost?

9. What will 9 quarts of milk cost, if 10 quarts cost 50 cents?

10. If a horse trot 36 miles in 6 hours, how many miles will he trot in 11 hours?

11. What will 12 yards of broadcloth cost, if 15 yards cost 75 dollars?

12. If a man can earn 90 dollars in 9 weeks, how many dollars can he earn in 6 weeks?

13. If 3 pounds of cheese cost 42 cents, what will 6 pounds cost?

---

* When any number or thing is divided into *four* equal parts, one of those parts is called *one fourth* ; when into *two* equal parts, one of those parts is called *one half* ; when into *three* equal parts, one of those parts is called *one third*, &c.

14. What cost 20 yards of cambric, if 6 yards cost 60 cents?

15. What cost 17 weeks' board, if 9 weeks' cost 27 dollars?

16. How many oranges can be bought for 63 cents, at the rate of 4 for 9 cents?

17. What cost 4 tons of coal, at the rate of 10 tons for 70 dollars?

18. If you should buy 60 pears, at the rate of 4 for 3 cents, and sell them at the rate of 5 for 4 cents, how much would you make by the operation?

19. When 56 cents are paid for 4 dozen of eggs, how much must be paid for 10 dozen?

20. How many pounds of veal can be bought for 96 cents, when 9 pounds cost 72 cents?

21. When 4 quarts of vinegar can be bought for 36 cents, how many quarts can be bought for 108 cents?

22. When 7 yards of cotton cloth can be bought for 84 cents, how many yards can be bought for 120 cents?

23. What will 19 pounds of chalk cost, if 13 pounds cost 65 cents?

24. When wheat is sold at the rate of 6 bushels for 12 dollars, how many bushels must be given for 4 cords of wood, at 5 dollars a cord?

25. If 8 dozen of eggs are sold for 96 cents, how many dozen will it take to buy 6 yards of gingham, at 18 cents a yard?

26. If it take 4 men 8 days to do a piece of work, how long will it take 16 men to do it?

27. How many men in 8 days can do a piece of work, which will require 16 men 2 days to do?

28. If 12 barrels of flour cost 84 dollars, what will 7 barrels cost? 11 barrels? 6 barrels? 10 barrels? 9 barrels?

29. If 9 men can dig a ditch in 8 days, how long will it take 12 men to dig it?

30. How many books, at 6 dimes each, will 9 books, at 4 dimes each, pay for?

31. A can travel at the rate of 5 miles an hour, and B 7 miles; they set out from the same point, and in the same direction, but B starts after A has travelled 30 miles; how long will it take B to overtake A?

32. William can run 40 rods in 5 minutes, and Jason the same distance in 4 minutes; how long will it take Jason to gain 30 rods on William?

33. Three men buy a horse, A paying 12 dollars, B twice as many lacking 4 dollars, and C twice as many as both of the others; what would have been the cost to each if the expense had been shared equally?

34. If 3 cords of wood are worth 18 dollars, and 10 cords are given for 12 thousand of shingles, how much are the shingles a thousand?

35. How much rye, at 9 dimes a bushel, must be given for 27 bushels of buckwheat, at 3 dimes a bushel?

36. If 2 bushels of oats are worth 1 bushel of corn, and 2 bushels of corn are worth 1 bushel of wheat, how many bushels of wheat are worth 20 bushels of oats?

37. If 6 men can reap a field in 4 days, in how many days can 8 men reap it?

38. When 3 pounds of cheese cost as much as 1 pound of butter, and 3 pounds of butter as much as 1 pound of tea, how many pounds of tea cost as much as 81 pounds of cheese?

39. If a cistern, capable of holding 60 gallons, has a pipe by which 10 gallons can run into it in one hour, and another pipe by which 5 gallons can run out of it in the same time, when both pipes are running in what time will the cistern be filled?

LESSON XIX.

TABLE OF UNITED STATES MONEY.

10 Mills	make 1 Cent,	c.
10 Cents	" 1 Dime,	d.
10 Dimes	" 1 Dollar,	$.
10 Dollars	" 1 Eagle,	E.

NOTE. — Dollars and cents written together are separated by a point (.); thus, $4.50 is read 4 dollars 50 cents.

1. How many mills in 1 cent? In 2 cents? In 9 cents? In 12 cents?

2. How many cents in 10 mills? In 20 mills? In 47 mills? In 80 mills?

3. How many cents in 1 dime? In 3 dimes? In 7 dimes? In 11 dimes? In 14 dimes?

4. How many dimes in 10 cents? In 20 cents? In 43 cents? In 90 cents? In 100 cents? In 110 cents? In 150 cents?

5. How many dimes in $1? In $3? In $7?

6. How many dollars in 1 eagle? In 2 eagles? In 7 eagles? In 11 eagles?

7. How many eagles in 10 dollars? In 20 dollars? In 70 dollars? In 120 dollars?

8. How many cents in $1? In $2? In $5? In $9?

9. How many dollars in 100 cents? In 200 cents? In 700 cents? In 900 cents?

10. At 5 mills a yard, how many cents will 20 yards of tape cost?

11. At 2 dimes a pound, how many dollars will 30 pounds of spice cost?

12. When butter is 2 dimes a pound, how many pounds will $5 buy?

13. If 7 pounds of beef cost 70 cents, how many pounds can be bought for 1 eagle?

TABLE OF ENGLISH MONEY.

4 Farthings (far.) make 1 Penny, d.
12 Pence " 1 Shilling, s.
20 Shillings " 1 Pound, £.

1. How many farthings in 1 penny? In 2 pence? In 6 pence? In 11 pence? In 20 pence?

2. How many pence in 4 farthings? In 12 farthings? In 36 farthings? In 48 farthings?

3. How many pence in 1 shilling? In 4 shillings? In 7 shillings? In 11 shillings?

4. How many shillings in 12 pence? In 48 pence? In 36 pence? In 108 pence?

5. How many shillings in 1£? In 4£? In 7£? In 9£? In 10£?

6. How many pounds in 20 shillings? In 40 shillings? In 80 shillings? In 120 shillings?

7. How many pence in 5s. 6d.? In 8s. 9d.? In 10s. 3d.?

8. How many shillings in 2£. 6s.? In 3£. 9s.? In 6£. 10s.?

9. At 3d. a pound, how many shillings will 24 pounds of rice cost?

10. At 5s. a yard, how many pounds will 32 yards of carpeting cost?

LESSON XX.

TABLE OF TROY, OR MINT WEIGHT.

24 Grains (gr.) make 1 Pennyweight, pwt.
20 Pennyweights " 1 Ounce, oz.
12 Ounces " 1 Pound, lb.

NOTE.—This weight is used in weighing gold, silver, and jewels.

1. How many grains in 1 pennyweight? In 2 pennyweights? In 4 pennyweights?

2. How many pennyweights in 24 grains? In 48 grains? In 96 grains?

3. How many pennyweights in 1 ounce? In 2 ounces? In 5 ounces?

4. How many ounces in 20 pennyweights? In 40 pennyweights? In 100 pennyweights?

5. How many ounces in 1 pound? In 2 pounds? In 5 pounds? In 9 pounds? In 10 pounds? In 12 pounds?

6. How many pounds in 12 ounces? In 36 ounces? In 60 ounces? In 144 ounces?

7. In 3oz. 10pwt., how many pennyweights?

8. In 6lb. 7oz., how many ounces?

9. At 6 cents a pennyweight, what will cost 4oz. 10pwt. of silver?

10. At 9 dimes a pennyweight, what must be paid for 2oz. 5pwt. of gold?

TABLE OF AVOIRDUPOIS WEIGHT.

16 Drams (dr.)	make 1 Ounce,	oz.
16 Ounces	" 1 Pound,	lb.
25 Pounds	" 1 Quarter,	qr.
4 Quarters	" 1 Hundred-weight,	cwt.
20 Hundred-weight	" 1 Ton,	T.

NOTE. — This weight is used in weighing almost every kind of goods, and all metals except gold and silver.

1. How many ounces in 1 pound? In 2 pounds? In 5 pounds? In 10 pounds?

2. How many pounds in 16 ounces? In 32 ounces? In 80 ounces? In 96 ounces?

3. How many pounds in 1 quarter? In 3 quarters? In 5 quarters? In 8 quarters?

4. How many quarters in 50 pounds? In 75 pounds? In 100 pounds?

5. How many hundred-weight in 4 quarters? In 8 quarters? In 16 quarters? In 20 quarters?

6. How many hundred-weight in 2 tons? In 3 tons? In 5 tons? In 6 tons?

7. How many tons in 20 hundred-weight? In 40 hundred-weight? In 80 hundred-weight? In 60 hundred-weight?

8. What cost 4 hundred-weight of sugar, at 9 cents a pound?

9. What cost 6 tons of bone-dust, at 2 dollars a hundred-weight?

10. How many pounds in 1cwt. 2qr. 13lb.?

11. How much will 2cwt. 1qr. of beef cost, at 10 cents a pound?

12. If 5cwt. of guano cost 15 dollars, how much will 3 tons cost?

LESSON XXI.

TABLE OF LINEAR, OR LONG MEASURE.

12 Inches (in.)	make 1 Foot,	ft.
3 Feet	" 1 Yard,	yd.
5½ Yards, or 16½ feet,	" 1 Rod or Pole,	rd.
40 Rods	" 1 Furlong,	fur.
8 Furlongs, or 320 rods,	" 1 Mile,	m.
3 Miles	" 1 League,	lea.
69⅙ Miles (nearly)	" 1 Degree,	deg. or °.
360 Degrees	" 1 Circle of the Earth.	

NOTE. — This measure is used in measuring distances in any direction.

½ is read one half, and ⅙ is read one sixth.

1. How many inches in 1 foot? In 5 feet? In 7 feet? In 10 feet? In 12 feet?

2. How many feet in 24 inches? In 36 inches?

3. How many feet in 2 yards? In 11 yards? In 15 yards? In 20 yards?

4. How many yards in 6 feet? In 12 feet? In 18 feet? In 24 feet?

5. How many rods in 1 furlong? In 3 furlongs? In 7 furlongs?

6. How many furlongs in 40 rods? In 120 rods? In 160 rods?

7. How many furlongs in 1 mile? In 3 miles? In 10 miles? In 12 miles?

8. How many miles in 16 furlongs? In 24 furlongs? In 40 furlongs? In 96 furlongs?

9. How many miles in 3 leagues? In 8 leagues? In 12 leagues? In 15 leagues?

10. How many leagues in 9 miles? In 24 miles? In 42 miles?

11. How many miles in 1 degree?

12. How many furlongs in 4 leagues? In 6 leagues?

13. If it take 5 minutes to travel 1 furlong, how long will it take to travel 1 mile? 1 league? 2 leagues?

14. How many inches in 2yd. 2ft. 6in.? In 1yd. 2ft. 5in.?

TABLE OF CLOTH MEASURE.

2¼ Inches	make	1 Nail,	na.
4 Nails	"	1 Quarter of a yard,	qr.
4 Quarters	"	1 Yard,	yd.
3 Quarters	"	1 Ell Flemish,	E. F.
5 Quarters	"	1 Ell English,	E. E.

NOTE. — This measure is used in measuring cloth, and other articles sold by the yard or ell.

¼ is read one fourth.

1. How many nails in 2 quarters? In 3 quarters?

2. How many quarters in 12 nails? In 16 nails? In 24 nails?

3. How many quarters in 1 yard? In 4 yards? In 12 yards? In 20 yards?

4. How many yards in 4 quarters? In 20 quarters? In 80 quarters? In 100 quarters?

5. How many quarters in 4 ells English? In 8 ells English? In 12 ells English?

6. How many ells English in 20 quarters? In 60 quarters? In 80 quarters?

7. What cost 10 yards of velvet, at $2 a quarter?

8. How many nails in 3yd. 3qr. 1na.?

9. How many quarters in 5yd. 3qr.?

10. 23 quarters are equal to how many yards?

11. At 3 cents a nail, how much will 3 ells English of cloth cost?

12. If 5 nails of cloth cost 25 cents, what cost 5 yards?

LESSON XXII.

TABLE OF SURFACE, OR SQUARE MEASURE.

144 Square inches (sq. in.) make	1 Square foot,	sq. ft.
9 Square feet	"	1 Square yard,	sq. yd.
30¼ Square yards	"	1 Square rod or pole,	p.
40 Square rods	"	1 Rood,	R.
4 Roods	"	1 Acre,	A.
640 Acres	"	1 Square mile,	S. M.

NOTE. — This measure is used in measuring surfaces of all kinds.

1. How many square feet in 1 square yard? In 3 square yards? In 7 square yards? In 8 square yards? In 9 square yards?

2. How many square yards in 9 square feet? In 36 square feet? In 81 square feet?

3. How many square rods in 1 rood? In 3 roods? In 7 roods?

4. How many roods in 80 square rods? In 120 square rods? In 160 square rods?

.5. How many roods in 2 acres? In 6 acres? In 10 acres?

6. How many acres in 8 roods? In 24 roods? In 60 roods?

7. How many acres in 1 square mile?

8. How many rods in a field 12 rods long, and 9 rods wide?

9. At 5 dollars for 1 square rod, what cost 1 acre of land?

10. In 1A. 2R. 20p. are how many square rods?

TABLE OF CUBIC, OR SOLID MEASURE.

1728 Cubic inches (cu. in.)	make 1 Cubic foot,	cu. ft.
27 " feet	" 1 " yard,	cu. yd.
40 " feet	" 1 Ton,	T.
16 " feet	" 1 Cord foot,	c. ft.
8 Cord feet, or } 128 Cubic feet, }	" 1 Cord of wood,	C.

NOTE. — This measure is used in measuring such things as have length, breadth, and thickness; as timber, stone, &c.

1. How many cubic feet in 1 cubic yard? In 3 cubic yards? In 4 cubic yards?

2. How many cubic yards in 27 cubic feet? In 54 cubic feet? In 81 cubic feet?

3. How many cubic feet in 2 tons of timber? In 3 tons of timber? In 5 tons of timber?

4. How many tons in 80 cubic feet of timber? In 120 cubic feet? In 160 cubic feet?

5. How many cord feet in 2 cords of wood? In 5 cords? In 9 cords?

6. How many cords in 16 cord feet? In 24 cord feet? In 80 cord feet?

7. How many cubic feet in 1 cord of wood? In 2 cords of wood?

8. What cost 5 cords of wood, if 4 cord feet cost $3?

9. What cost 2 tons 20 cubic feet of timber, at $1 for 4 cubic feet?

LESSON XXIII.

LIQUID, OR WINE MEASURE.

4 Gills (gi.)	make 1 Pint,	pt.
2 Pints	" 1 Quart,	qt.
4 Quarts	" 1 Gallon,	gal.
63 Gallons	" 1 Hogshead,	hhd.
2 Hogsheads	" 1 Pipe,	pi.
2 Pipes	" 1 Tun,	tun.

NOTE.—This measure is used in measuring all kinds of liquid, except, in some places, beer, ale, porter, and milk.

1. How many gills in 4 pints? In 8 pints? In 12 pints?

2. How many pints in 8 gills? In 16 gills? In 32 gills?

3. How many pints in 4 quarts? In 6 quarts? In 12 quarts?

4. How many quarts in 3 gallons? In 8 gallons? In 15 gallons?

5. How many gallons in 1 pipe?

6. How many hogsheads in 2 tuns? In 6 tuns?

7. How many tuns in 8 hogsheads? In 16 hogsheads? In 20 hogsheads?

8. What cost 8 gallons of vinegar, at 5 cents a quart? At 8 cents? At 10 cents?

9. If 2 quarts of oil cost 48 cents, what cost 1 quart? 1 pint? 1 gill?

10. If 2 gills of molasses cost 4 cents, how much will 2 gallons cost?

11. If 3 quarts of oil cost 60 cents, what cost 3 gallons and 3 quarts?

12. How many gills in 2 quarts and 1 pint?

13. When 5 gallons of burning-fluid can be bought for 3 dollars, what cost 1 hogshead and 7 gallons?

TABLE OF DRY MEASURE.

2 Pints   make 1 Quart,   qt.
8 Quarts   "   1 Peck,   pk.
4 Pecks   "   1 Bushel,   bu.

NOTE. — This measure is used in measuring grain, fruit, salt, coal, &c.

1. How many pints in 4 quarts?   In 6 quarts? In 10 quarts?

2. How many quarts in 12 pints?   In 18 pints? In 20 pints?   In 30 pints?

3. How many quarts in 4 pecks?   In 8 pecks?

4. How many pecks in 16 quarts?   In 32 quarts?   In 64 quarts?

5. How many pecks in 4 bushels?   In 5 bushels?   In 9 bushels?

6. How many bushels in 12 pecks?   In 32 pecks?   In 48 pecks?

7. What costs 1 bushel of corn, at 3 cents a quart?

8. What cost 2 pecks of cherries, at 4 cents a pint?

9. How many quarts in 2 bushels?   In 3 bushels?

10. How many pints in 2bu. 3pk. 4qt.?

11. If 3 pints of cherries cost 15 cents, what will 1 bushel cost?

LESSON XXIV.

TABLE OF TIME.

60 Seconds (sec.)   make 1 Minute,   m.
60 Minutes   "   1 Hour,   h.
24 Hours   "   1 Day,   d.
7 Days   "   1 Week,   w.
365¼ days, or 52 weeks, 1¼ days,   "   1 Julian Year, y.
12 Calendar months   "   1 Year.

Note. — This measure is applied to the various divisions and subdivisions into which time is divided.

The following table will exhibit the names of the months, and the number of days in each.

1st	month,	January,	has	31	days.
2nd	"	February,	"	28	"
3d	"	March,	"	31	"
4th	"	April,	"	30	'
5th	"	May,	"	31	"
6th	"	June,	"	30	"
7th	"	July,	"	31	"
8th	"	August,	"	31	"
9th	"	September,	"	30	"
10th	"	October,	"	31	"
11th	"	November,	"	30	"
12th	"	December,	"	31	".

{ except in Leap-year, when it has 29.

1. How many seconds in 2 minutes? In 3 minutes?

2. How many minutes in 120 seconds?

3. How many minutes in 2 hours? In 4 hours?

4. How many hours in 2 days? In 3 days?

5. How many days in 48 hours? In 72 hours?

6. How many days in 4 weeks? In 6 weeks? In 9 weeks? In 10 weeks?

7. How many weeks in 14 days? In 63 days? In 84 days?

8. How many months in 6 years? In 10 years?

9. If you can read 6 pages in 12 minutes, how many hours will it take you to read 10 times as many?

10. Charles is 8 years 3 months old, and John is 7 years 10 months; how many months is Charles older than John?

11. If a man can earn 60 dollars in 3 months, in how many months can he earn 100 dollars?

12. If in 5 hours 3 pairs of shoes can be made, in how many days, of 10 hours each, can 24 pairs be made?

MISCELLANEOUS TABLE.

12 Units	make 1 Dozen.
12 Dozen	" 1 Gross.
12 Gross	" 1 Great Gross.
20 Units	" 1 Score.
24 Sheets of paper	" 1 Quire.
20 Quires	" 1 Ream.
56 Pounds	" 1 Bushel of Corn.
60 Pounds	" 1 Bushel of Wheat.
196 Pounds	" 1 Barrel of Flour.
200 Pounds	" 1 Barrel of Beef.
200 Pounds	" 1 Barrel of Pork.

1. How many units in 4 dozen? In 8 dozen?

2. What cost 4 dozen peaches, at 2 cents apiece?

3. What costs 1 gross of writing-books, at 10 cents each?

4. How many score in 60 pounds? In 100 pounds?

5. What cost 12 score of pork, at 10 cents a pound?

6. What costs 1 gross of pens, at 5 cents a dozen?

7. What costs 1 ream of paper, at 10 cents a quire?

8. What costs 1 quire of paper, when 3 sheets can be bought for 2 cents?

9. At $11 a hundred-weight, what costs 1 barrel of beef?

10. Bought a barrel of pork at 12 cents a pound, and sold it at $15 a hundred-weight; how much was made by the sale?

11. Bought wheat at 3 cents a pound, and sold it at $2 a bushel; how much was made on a bushel?

12. Bought beef at $18 a barrel, and sold it at 12 cents a pound; how much was made on a barrel?

LESSON XXV.

1. When any thing or number is divided into two equal parts, what is each one of those parts called?   Ans. One half.

2. How many halves in a whole one?

3. What is meant by a half of any thing?

4. If a pear be worth 2 cents, how much of it is worth 1 cent?

5. What part of 2 is 1?

6. How many halves are there in 2?

NOTE. — Since there are 2 halves in 1, in 2 there are 2 times 2 halves, which are 4 halves.

7. How many halves are there in 4? In 5? In 10? In 30? In 50?

8. How many halves in 2 and 1 half? Ans. 5 halves.

9. How many halves in 3? In 4 and 1 half? In 9 and 1 half? In 11 and 1 half? In 12 and 1 half?

10. How many whole ones are there in 2 halves? In 6 halves? In 30 halves?

11. What is 1 half of 3? Ans. 1 and 1 half.

12. What is 1 half of 5? Of 7? Of 19? Of 6? Of 32? Of 55?

13. 7 is how many times 2?

14. 13 is how many times 2?

15. When any thing or number is divided into three equal parts, what is each one of those equal parts called?   Ans. One third.

16. What are two of those equal parts called? Ans. Two thirds.

17. How many thirds in a whole one?

18. What is meant by one third of any thing? By two thirds?

19. If 1 third of an orange be worth 1 cent, what are 2 thirds of it worth? What is the whole of it worth?

20. If an orange be worth 3 cents, how much of it is worth 1 cent? 2 cents?

21. How many thirds are there in 2? In 3? In 6? In 9? In 10?

22. How many thirds in 2 and 1 third? In 4 and 2 thirds? In 11 and 1 third?

23. How many whole things in 6 thirds? In 7 thirds? In 19 thirds?

24. What is 1 third of 4? Ans. 1 and 1 third.

25. 4 is how many times 3?

26. 9 is how many times 3?

27. 16 is how many times 3?

28. 47 is how many times 3?

29. How many is 1 third of 5? Of 11? Of 17?

30. When a cord of wood costs $6, what costs 1 third of a cord? 2 thirds of a cord?

Note. — Since 1 cord costs $6, 1 third of a cord costs 1 third part of $6, which is $2; two thirds cost 2 times $2, which are $4.

31. If a barrel of sugar is worth $18, what is 1 third of a barrel worth? 2 thirds of a barrel?

32. When 5 quarts of cherries can be bought for 35 cents, what part of 21 cents will 1 quart cost? 2 quarts?

33. If an apple be divided into 4 equal parts, what is one of those parts called? Ans. One fourth.

34. What are two of those parts called? Three of those parts called?

35. How many fourths in a whole one?

36. What is meant by a fourth of any thing? By two fourths? By three fourths?

37. What part of 4 is 1? 2? 3?

38. How many fourths in 2?   In 3 and 1 fourth?
39. 5 is how many times 4?
40. 7 is how many times 4?
41. 10 is how many times 4?
42. 17 is how many times 4?
43. What is 1 fourth of 16?   21?   28?   39?
56?

44. If 4 dimes will purchase a yard of flannel, what part of a yard will 1 dime purchase?   2 dimes?   3 dimes?

45. If a bushel of corn cost 60 cents, what will one fourth of a bushel cost?   Two fourths?   Three fourths?

46. If 3 bushels of cranberries cost $12, what part of $16 will 1 bushel cost?

47. James has 48 apples, and his brother has 3 fourths as many; how many has his brother?

48. Bought a horse for $160, and a wagon for $5 more than 3 fourths the cost of the horse; what was the cost of the wagon?

LESSON XXVI.

1. If a barrel of flour cost $10, what part of a barrel will cost $2?

2. What is meant by 1 fifth of any thing?   By 2 fifths?   By 3 fifths?   By 4 fifths?

3. How many fifths in a whole one?

4. What part of 5 is 3?   2?   4?

5. 8 is how many times 5?

6. 9 is how many times 5?

7. 29 is how many times 5?

8. 47 is how many times 5?

9. If a piece of land is worth $75, how many times $5 is 1 fifth of it worth?

10. Harriet had 40 pins, and gave Maria 1 fifth of them; how many did she give her?

11. If 13 pounds of veal cost 91 cents, what part of 42 cents will 2 pounds cost?

12. How many sixths in a whole one? In 3? In 2? In 10? In 11?

13. What is meant by 1 sixth of any thing? By 2 sixths? By 5 sixths?

14. What part of 6 is 1? 2? 3? 4?

15. What is one sixth of 6? Of 12? Of 18? Of 31? Of 93?

16. When a pound of spice can be bought for 24 cents, what part of a pound can be bought for 4 cents?

17. At $7 a yard, what costs 1 sixth of a yard of broadcloth?

18. At the rate of $6 a cord, how much wood can be bought for $1? For $5?

19. 7 is how many times 6?

20. 15 is how many times 6?

21. 24 is how many times 6?

22. 35 is how many times 6?

23. 81 is how many times 6?

24. Bought a horse for 120 dollars, and sold him at 1 sixth more than he cost; for how much was he sold?

25. If strawberries are 6 cents a pint, how many quarts can be bought for 62 cents?

26. How many whole ones in 17 sixths?

27. How many whole ones in 37 sixths?

28. How many whole ones in 73 sixths?

29. How many sevenths in a whole one?

30. What is meant by 1 seventh of any thing? By 2 sevenths? By 4 sevenths? By 6 sevenths?

31. When 35 cents will buy a pound of butter, how many cents will buy 1 seventh of a pound? 3 sevenths of a pound? 5 sevenths of a pound?

32. If $14 will buy a barrel of beef, how much of a barrel will $2 buy? $4? $8? $10? $12?

33. How many whole ones in 14 sevenths? In 17 sevenths? In 23 sevenths? In 39 sevenths? In 84 sevenths?

34. 17 is how many times 7?

35. 46 is how many times 7?

36. 35 is how many times 7?

37. 78 is how many times 7?

38. If a man can earn $42 in a month, how many dollars can he earn in 1 seventh of a month? In 2 sevenths? In 4 sevenths? In 6 sevenths?

39. A boy at recitation having answered 4 questions wrong, found the number to be just 2 sevenths of all the questions asked; how many were asked?

40. What is meant by 1 eighth of any thing? By 3 eighths?

41. How many ninths in a whole one? How many tenths?

42. 64 is how many times 9?

43. 78 is how many times 8?

44. 93 is how many times 10?

45. 100 is how many times 11?

46. 105 is how many times 12?

47. Edward has 48 walnuts, and Alfred has 5 eighths as many; how many has Alfred?

48. What cost 3 tenths of a ton of hay, at 7 dimes a hundred-weight?

49. William earns 63 cents a day, and John 10 ninths as many and 5 cents more; how many does John earn?

50. What cost 2 ninths of a hogshead of molasses, at 10 cents a quart?

51. What cost 5 twelfths of a gross of writing-books, at 10 cents a book?

52. Robert, having $60 to spend, paid 1 twelfth of it for a hat, 2 fifths for a suit of clothes, and 3 tenths for tuition; how much then had he left?

LESSON XXVII.

1.   8 is 1 fourth of what number?
2.   8 is 1 third of what number?
3.   7 is 1 fifth of what number?
4.   9 is 1 fourth of what number?
5.   12 is 1 third of what number?
6.   15 is 1 sixth of what number?
7.   18 is 1 tenth of what number?
8.   6 is 3 fourths of what number?
9.   12 is 2 fifths of what number?
10.   8 is 4 fifths of what number?
11.   9 is 3 sixths of what number?
12.   10 is 5 eighths of what number?
13.   16 is 4 ninths of what number?
14.   18 is 9 tenths of what number?
15.   20 is 4 fifths of what number?
16.   24 is 6 eighths of what number?
17.   30 is 3 elevenths of what number?
18.   14 is 7 eighths of how many times 4?
19.   9 is 3 fourths of how many times 6?
20.   10 is 5 sixths of how many times 4?
21.   24 is 3 ninths of how many times 12?
22.   20 is 4 fifths of how many times 5?
23.   24 is 4 fifths of how many times 10?
24.   18 is 3 ninths of how many times 6?
25.   30 is 3 fifths of how many times 10?
26.   28 is 4 sevenths of how many times 7?
27.   24 is 4 twelfths of how many times 9?
28.   15 is 1 fourth of 5 times what number?
29.   18 is 1 third of 6 times what number?
30.   12 is 1 sixth of 8 times what number?
31.   22 is 1 half of 11 times what number?
32.   30 is 1 third of 15 times what number?
33.   16 is 1 seventh of 8 times what number?
34.   9 is 1 ninth of 9 times what number?

35. 36 is 1 tenth of 20 times what number?
36. 5 eighths of 32 are how many times 5?
37. 6 eighths of 48 are how many times 12?
38. 7 ninths of 54 are how many times 6?
39. 4 tenths of 70 are how many times 7?
40. 3 elevenths of 44 are how many times 6?
41. 5 twelfths of 84 are how many times 7?
42. 8 sevenths of 63 are how many times 8?
43. 4 times 3 and 2 thirds of 3 are how many?
44. 5 times 4 and 3 fourths of 4 are how many?
45. 7 times 6 and 4 sixths of 6 are how many?
46. 8 times 4 and 3 fourths of 4 are how many?
47. 5 times 7 and 4 sevenths of 7 are how many?
48. 9 times 8 and 7 eighths of 8 are how many?
49. 6 times 9 and 5 ninths of 9 are how many?
50. 5 times 9 and 3 ninths of 9 are how many?
51. 8 times 10 and 9 tenths of 10 are how many times 5?
52. 7 times 7 and 5 sevenths of 7 are how many times 6?
53. 9 times 12 less 11 twelfths of 12 are how many times 10?
54. 10 times 11 less 7 elevenths of 11 are how many times 8?
55. 2 thirds of 9 is 1 fourth of what number?
56. 3 fourths of 12 is 1 third of what number?
57. 4 fifths of 15 is 3 fifths of what number?
58. 5 ninths of 36 is 2 fifths of how many times 10?
59. 4 tenths of 30 is 2 thirds of how many times 3?
60. 5 sixths of 18 is 3 fifths of how many times 5?
61. 6 tenths of 20 is 1 seventh of what number?
62. 5 twelfths of 60 is how many times 1 sixth of 30?

LESSON XXVIII.

1. If a melon costs 8 cents, what will 1 fourth of a melon cost?

2. At $5 a yard, what will 1 fifth of a yard cost?

3. At $6 a ton, what will 7 sixths of a ton cost?

4. At $12 a barrel, what will 1 barrel and 1 sixth of a barrel of flour cost?

5. When a man is working for $15 a week, how much does he earn in 2 weeks and 1 fifth of a week?

6. At 16 cents a pound, what cost 3 pounds and 3 fourths of a pound of sugar?

7. At $8 a cord, what cost 4 cords and 3 fourths of a cord of wood?

8. At 6 dimes a bushel, what cost 10 bushels and 2 thirds of a bushel of corn?

9. A teacher, being asked how many scholars he had, replied that the smallest number that ever had been present was 18, which was just 3 ninths of his whole number; how many had he?

10. Sold a horse for $150, which was 5 thirds of what he cost; what did he cost?

11. I have $4, which is 1 fifth of what I divided among 10 men; how much did each receive?

12. William has 63 cents in his pocket, which is 9 fifths of what he has in his money-box; how much has he in his money-box?

13. A boy, being asked how many chickens he had, answered that his largest brood contained 20, which was 4 fifths of the whole number, and the whole number was 5 times as many as he had in the smallest brood; how many had he, and how many were there in the smallest brood?

14. A farmer had 42 sheep, and sold 5 sevenths of them to 5 of his neighbors, each receiving an equal number; how many did each receive?

15. I had 24 cherries, but have divided 5 sixths of them among 10 children; how many does each receive?

16. A gentleman had 36 pears, and gave 1 sixth of them to Frank, and divided the remainder equally among his 5 brothers; how many did Frank receive more than each of his brothers?

17. At 10 cents a pound, what cost 20 pounds and 3 tenths of a pound of raisins?

18. At 18 cents a pound, what cost 8 pounds and 5 ninths of a pound of butter?

19. Henry gave $5 for a vest, which was just 1 half of 4 times as much as he gave for a hat; how much did he give for the hat?

20. What cost 5 barrels of flour, if 1 tenth of a barrel cost 9 dimes?

21. John has 20 cents, which is 4 fifths of 5 times as many as Joseph has; how many has Joseph?

22. Emma is 12 years old, and 7 sixths of her age is just 7 eighths of her sister's age; what is her sister's age?

23. A pier of a certain bridge stands 10 feet in the water, which is 2 fifths of the height of the pier lacking 5 feet; what is the height of the pier?

24. Bought 4 pounds and 3 fourths of a pound of rice at 8 cents a pound, and paid for it with berries at 5 cents a quart; how many quarts did it take?

25. Bought a watch for $40, and a chain for $5 more than 2 fifths the cost of the watch; what did both cost?

26. Sold a hogshead of molasses for $36, which was 9 eighths of what it cost; what did it cost?

27. A farmer has 6 cows, and 2 thirds the number of cows is equal to 1 fourth the number of his sheep; how many sheep has he?

28. Four boys wished their father to divide 50 cents equally among them. "No," said he; "but I will give you 4 fifths of it, and one half of the remaining fifth to the one that will tell how to divide it;" how much would the one who could tell receive?

29. If 6 pounds of sugar cost 96 cents, what cost 5 eighths of a pound?

30. A man, wishing to draw the water from a well containing 120 gallons, found that while he could draw out 12 gallons in a minute, 2 thirds as many gallons would run in during the same time; how long would it take him to exhaust the well?

31. Bought 9 yards of broadcloth for $36, which was 4 fifths of its real value; what was its real value per yard?

32. James, in reciting, had given in one day 8 imperfect answers, and 3 fourths of the imperfect answers were just 2 thirds of the number of perfect answers; how many perfect answers had he given?

33. In a small basket there are 12 pears, and 2 thirds of these are 2 sixths of the number in a large basket; if the whole be divided among 4 boys, how many will each receive?

34. A pole stands 3 fifths in the water, 1 half of the remainder in the mud, and 4 feet above the water; what is the length of the pole?

35. Frank had 2 bags filled with nuts, and his brother wished to buy 1 of them. "William," said Frank, "I have 9 quarts in the smaller bag, and 2 thirds of that number is just 1 half the number of quarts in the larger bag; and if you will

6

tell me the number of quarts in the larger bag, I will give them to you;" how many were there in both bags?

36. George spent 6 elevenths of his money for a suit of clothes; he then paid $2 for a hat, which was just 1 fifth of all he had left; how many dollars had he at first?

37. A certain school is divided into three classes; in the first class there are 54 pupils, which is 9 tenths of the number in the second class, and the number in the second class is equal to 6 fifths of 2 times the number in the third class; how many are there in each of the classes?

LESSON XXIX.

Such expressions as one half, one third, two thirds, &c., are called fractions, and denote one or more equal parts of a unit. Fractions are expressed thus:

$\frac{1}{2}$ one half.		$\frac{3}{5}$ three fifths.	
$\frac{1}{3}$ one third.		$\frac{4}{5}$ four fifths.	
$\frac{2}{3}$ two thirds.		$\frac{1}{6}$ one sixth.	
$\frac{1}{4}$ one fourth.		$\frac{5}{6}$ five sixths.	
$\frac{2}{4}$ two fourths.		$\frac{3}{7}$ three sevenths.	
$\frac{3}{4}$ three fourths.		$\frac{9}{8}$ nine eighths.	
$\frac{1}{5}$ one fifth.		$\frac{11}{12}$ eleven twelfths.	
$\frac{2}{5}$ two fifths.		$\frac{13}{19}$ thirteen nineteenths.	

NOTE. — The figure or figures above the short horizontal line are called the numerator of the fraction, and the figure or figures below the line are called the denominator. The denominator shows into how many equal parts a unit has been divided, and the numerator shows how many of these equal parts have been taken.

When the numerator is less than the denominator, the fraction is called a *proper fraction;* when the numerator is equal to or larger than the denominator, the fraction is called an *improper fraction;* when a fraction is joined with a whole number, the expression is called a *mixed number.*

1. What kind of a fraction is $\frac{1}{2}$? What is the 2 called? What does it show?

2. Into how many equal parts is the unit divided, to give the fraction $\frac{1}{3}$? How many of these equal parts are taken?

3. What kind of a fraction is $\frac{4}{7}$? Why?

4. In the fraction $\frac{2}{3}$, what is the 2 called? What does it show?

5. In the fraction $\frac{7}{9}$, what is the 9 called? Into how many equal parts does it show the thing to be divided?

6. What kind of a fraction is $\frac{4}{3}$? Why?

7. In the fraction $\frac{7}{8}$, what does the 7 show? What does the 8 show?

8. What is the expression $5\frac{2}{11}$ called? Why?

9. Into how many equal parts is the unit divided to give the fraction $\frac{1}{4}$? How many of the parts are taken?

10. How can you find $\frac{3}{5}$ of any thing?

11. How can you find $\frac{2}{9}$ of any thing?

12. In 12 how many fourths?

13. Reduce 15 to thirds. To fifths.

14. How can you reduce a whole number to a fraction having any given denominator?

15. How many times $\frac{1}{4}$ in 5? In 16?

16. How many times $\frac{1}{6}$ in 7? In 10?

17. How many times $\frac{1}{4}$ in 9? In 12?

18. In $6\frac{2}{3}$ how many thirds?

19. Reduce $7\frac{2}{5}$ to an improper fraction.

Note. —That is, change 7 and 2 fifths to fifths.

20. Reduce $9\frac{5}{6}$ to an improper fraction.

21. Express $12\frac{1}{4}$ by an improper fraction.

22. Express $18\frac{5}{10}$ by an improper fraction.

23. How do you change a mixed number to an improper fraction?

24. How many times 1 in $\frac{8}{4}$? In $\frac{11}{4}$?

25. How many times 1 in $\frac{44}{11}$? In $\frac{38}{19}$?
26. How many times 1 in $\frac{39}{13}$? In $\frac{54}{11}$?
27. How many times 1 in $\frac{44}{11}$? In $\frac{56}{14}$?
28. How many times 1 in $\frac{63}{9}$? In $\frac{72}{9}$?
29. How many times 1 in $\frac{108}{12}$? In $\frac{99}{11}$?
30. How many times 1 in $\frac{126}{10}$? In $\frac{84}{10}$?
31. How many times 1 in $\frac{96}{4}$? In $\frac{33}{4}$?
32. How many times 1 in $\frac{82}{2}$? In $\frac{102}{3}$?
33. How many times 1 in $\frac{35}{3}$? In $\frac{66}{3}$?
34. How many times 1 in $\frac{48}{2}$? In $\frac{84}{4}$?
35. How many times 1 in $\frac{101}{4}$? In $\frac{64}{4}$?
36. Reduce $\frac{32}{8}$ to an equivalent whole number.
37. Reduce $\frac{8}{5}$ to an equivalent mixed number.
38. Reduce $\frac{42}{4}$ to an equivalent mixed number.
39. Reduce $\frac{17}{3}$ to an equivalent mixed number.
40. Reduce $\frac{25}{8}$ to an equivalent mixed number.
41. Reduce $\frac{88}{11}$ to an equivalent whole number.
42. Reduce $\frac{104}{2}$ to an equivalent whole number.
43. Reduce $\frac{175}{25}$ to an equivalent whole number.
44. Express $\frac{99}{11}$ by an equivalent whole number.
45. Express $\frac{100}{24}$ by an equivalent mixed number.
46. Express $\frac{163}{40}$ by an equivalent mixed number.
47. Express $\frac{148}{12}$ by an equivalent mixed number.
48. Express $\frac{131}{10}$ by an equivalent mixed number.
49. Express $\frac{126}{60}$ by an equivalent mixed number.
50. How do you reduce an improper fraction to an equivalent whole or mixed number?
51. Reduce $\frac{4}{8}$ to its lowest terms. Ans. $\frac{1}{2}$.

NOTE. — A fraction is in its lowest terms when no number greater than 1 will divide both its numerator and denominator, without a remainder. *Dividing the numerator and denominator of a fraction by the same number, does not alter the value of the fraction.*

52. Reduce $\frac{6}{12}$ and $\frac{9}{15}$ to their lowest terms.
53. Reduce $\frac{5}{10}$ and $\frac{4}{12}$ to their lowest terms.
54. Reduce $\frac{7}{14}$ and $\frac{5}{15}$ to their lowest terms.
55. Reduce $\frac{6}{18}$ and $\frac{8}{24}$ to their lowest terms.
56. Reduce $\frac{8}{27}$ and $\frac{8}{32}$ to their lowest terms.
57. Reduce $\frac{14}{28}$ and $\frac{24}{48}$ to their lowest terms.
58. Reduce $\frac{15}{60}$ and $\frac{9}{72}$ to their lowest terms.
59. Reduce $\frac{12}{36}$ and $\frac{42}{84}$ to their lowest terms.
60. Reduce $\frac{50}{80}$ and $\frac{25}{100}$ to their lowest terms.
61. Reduce $\frac{60}{75}$ and $\frac{75}{200}$ to their lowest terms.
62. Reduce $\frac{24}{64}$ and $\frac{56}{63}$ to their lowest terms.
63. When is a fraction in its lowest terms?
64. How do you reduce a fraction to its lowest terms?
65. Why is the value of a fraction not changed by reducing it to its lowest terms?

LESSON XXX.

1. Gave $\frac{2}{4}$ of an apple to one boy, $\frac{3}{4}$ to another, and $\frac{3}{4}$ to another; how many fourths were given away? How many whole apples?
2. Lydia has $\frac{3}{6}$ of a dollar, and Mary $\frac{5}{6}$ of a dollar; how many dollars have they both?
3. Sold $\frac{5}{6}$ of an acre of land to one man, $\frac{5}{6}$ to another, and $\frac{2}{6}$ to another; how many acres were sold?
4. $\frac{4}{6} + \frac{2}{6} + \frac{5}{6} + \frac{4}{6}$ are how many times 1?
5. Edmund had $7\frac{3}{4}$ dollars, and his father gave him $\frac{3}{4}$ of a dollar more; how many dollars then had he?

NOTE. $\frac{3}{4} + \frac{3}{4} = \frac{6}{4} = \frac{3}{2} = 1\frac{1}{2}$; and $7 + 1\frac{1}{2} = 8\frac{1}{2}$.

6. Bought a barrel of flour for $$9\frac{5}{8}$, and a yard of velvet for $$4\frac{7}{8}$; how much did the whole cost?
7. James gathered $8\frac{5}{8}$ quarts of berries, Frank

$4\frac{5}{8}$ quarts, and Arthur $6\frac{1}{8}$ quarts; how many did they all gather?

8. $6\frac{3}{7} + 11\frac{1}{7} + 10\frac{5}{7}$ are how many times 1?

9. $3\frac{1}{5} + 4\frac{3}{5} + \frac{4}{5}$ are how many times 1?

10. $8\frac{3}{8} + 4\frac{7}{8} + 4$ are how many times 1?

11. $\frac{11}{12} + \frac{7}{12} + 19$ are how many times 1?

12. $5\frac{3}{10} + 7\frac{7}{10} + 12\frac{8}{10}$ are how many times 1?

13. If you should give $\frac{4}{8}$ of a melon to your brother, and retain the rest for yourself, what part would you retain?

NOTE. $1 = \frac{8}{8}$; and $\frac{8}{8} - \frac{4}{8} = \frac{4}{8}$.

14. Sarah bought 3 yards of cloth, and gave Ellen $\frac{3}{4}$ of a yard; how much had she left?

15. Bought a bible for $5, less $\frac{7}{8}$; how much was paid for it?

16. A gentleman owned a ship, but has sold $\frac{5}{16}$ of it; what part does he still own?

17. George had $17\frac{2}{5}$, but has spent $2\frac{4}{5}$ of it; how much has he left?

NOTE. $17\frac{2}{5} - 2 = 15\frac{2}{5} = 14\frac{7}{5}$; and $14\frac{7}{5} - \frac{4}{5} = 14\frac{3}{5}$.

18. From $25\frac{7}{12}$ acres of land there have been sold $4\frac{11}{12}$ acres; how much is left?

19. From a hogshead of wine there leaked out $6\frac{3}{4}$ gallons; how many gallons remained?

20. $75 - 12\frac{4}{5}$ are how many times 1?

21. $93\frac{1}{7} - 10\frac{6}{7}$ are how many times 1?

22. $82\frac{1}{8} - 4\frac{5}{8}$ are how many times 1?

23. George put into the bank at one time $11, and at another time $6\frac{5}{12}$; how much more must he put in to make up $20?

24. Three men bought a boat together; A paid $18\frac{4}{10}$, C $20\frac{7}{10}$, and B the balance; how much did B pay, if the boat cost $50?

25. A farmer had 60 bushels of apples, but sold

at one time $16\frac{3}{8}$ bushels, and at another time $30\frac{7}{8}$ bushels; how many had he left?

26. $10 + 6\frac{2}{3} - 9$ are how many?

27. $20\frac{4}{5} + 10\frac{3}{5} - 6\frac{3}{5}$ are how many?

28. $21\frac{4}{11} + 7\frac{8}{11} - 10\frac{1}{11}$ are how many?

29. $16\frac{1}{3} + 8\frac{2}{3} - 20\frac{1}{3}$ are how many?

30. Bought a coat for \$$11\frac{5}{8}$, and a vest for \$$3\frac{7}{8}$, and gave in payment two ten-dollar bills; how much change should be received back?

31. George is $11\frac{3}{12}$ years old, Simeon $8\frac{7}{12}$, and Edward $6\frac{3}{12}$; how much does the sum of their ages exceed 5 times 5?

32. Sold 6 cords of wood, at \$5 a cord, and bought a hundred-weight of sugar for \$$14\frac{3}{4}$, and a thousand feet of boards for \$$15\frac{3}{4}$; how much did the articles bought cost more than was received for the wood?

33. Sold a cow for \$40, and took in part payment a plough worth \$$11\frac{2}{3}$, and a harness worth \$$20\frac{2}{3}$; how much was then due?

34. Bought 3 barrels of beef, at \$14 a barrel, and paid in merchandise \$$24\frac{7}{10}$, and the balance in cash; how much less was paid in cash than in merchandise?

LESSON XXXI.

1. If a family consume $\frac{2}{5}$ of a barrel of flour in 1 week, how much will it consume in 6 weeks?

> NOTE. 6 times $\frac{2}{5}$ are $\frac{12}{5} = 2\frac{2}{5}$.

2. What cost 7 yards of cloth, at $\frac{4}{9}$ of a dollar a yard?

3. At $\frac{2}{3}$ of a cent apiece, what cost 12 eggs?

4. At $\frac{3}{4}$ of a dollar a day, how much can be earned in 12 days?

5. If a man can walk $3\frac{1}{2}$ miles in an hour, how far can he walk in 10 hours?

6. What cost 7 chairs, at $5\frac{1}{2}$ dollars apiece?

7. At $\frac{1}{8}$ of a dollar a peck, what cost 5 pecks of apples? 7 pecks? 8 pecks?

8. At $\frac{7}{16}$ of a dollar a bushel, what cost 3 bushels of potatoes? 8 bushels? 12 bushels?

9. How many are 4 times $\frac{1}{8}$? 5 times $\frac{4}{5}$?

10. How many are 6 times $\frac{3}{5}$? 7 times $\frac{2}{7}$?

11. How many are 9 times $\frac{1}{4}$? 10 times $\frac{8}{9}$?

12. At $6\frac{1}{4}$ cents a pound, what cost 9 pounds of rice? 10 pounds?

13. When eggs are $16\frac{2}{3}$ cents a dozen, what cost 6 dozen?

14. If a peck of corn cost $\frac{3}{16}$ of a dollar, how much will 2 bushels cost?

15. What cost $14\frac{3}{4}$ pounds of cheese, at 10 cents a pound?

16. What cost 20 bushels of wheat, at $\$2\frac{1}{5}$ a bushel?

17. What cost 13 yards of silk, at $\$2\frac{3}{4}$ a yard?

18. At $6\frac{1}{4}$ cents a nail, what cost 8 yards of cloth? 10 yards?

19. How many are 4 times $2\frac{1}{2}$? 6 times $3\frac{2}{5}$?

20. How many are 6 times $5\frac{2}{3}$? 9 times $5\frac{3}{8}$?

21. How many are 8 times $4\frac{1}{2}$? 10 times $5\frac{3}{8}$?

22. How many are 7 times $8\frac{7}{8}$? 12 times $11\frac{4}{12}$?

23. If a horse trot $9\frac{3}{4}$ miles in 1 hour, how far, at that rate, can he trot in 9 hours?

24. How much can be earned in a year, at $\$11\frac{1}{3}$ a month?

25. What cost 5 bushels of corn, at $\frac{4}{5}$ of a dollar a bushel?

NOTE. — Dividing the denominator by any number affects the value of a fraction the same as multiplying the numerator. The denominator of $\frac{4}{5}$ divided by 5 gives $\frac{4}{1}=4$.

26. If a man can reap $\frac{7}{8}$ of an acre in a day, how many acres can 4 men reap in the same time?

27. What cost 6 pounds of opium, at $\$4\frac{1}{8}$ a pound?

28. How many are 8 times $5\frac{3}{16}$?   8 times $12\frac{1}{2}$?

29. How many are 9 times $10\frac{1}{2}$?   9 times $2\frac{7}{18}$?

30. How many are 10 times $10\frac{3}{20}$?   10 times $15\frac{9}{10}$?

31. How many are 9 times $\frac{34}{63}$?   9 times $\frac{72}{54}$?

32. How many times 11 are 6 times $\frac{41}{12}$?

33. How many times 25 are 21 times $\frac{300}{63}$?

34. How do you multiply a fraction by a whole number?

35. How many cords of wood, at $\$5$ a cord, will pay for 8 barrels of flour, at $\$10\frac{1}{2}$ a barrel?

36. How many loads of hay, at $\$11$ a load, will pay for 5 yards of cloth, at $\$4\frac{9}{10}$ a yard, and a debt of $\$8\frac{5}{10}$?

LESSON XXXII

1. At $\frac{4}{9}$ of a dollar a yard, how many yards of cloth can be bought for $\$3\frac{1}{9}$?

NOTE. $3\frac{1}{9} = \frac{28}{9}$; and $\frac{28}{9}$ divided by $\frac{4}{9} = 28$ divided by $4 = 7$.

2. At $\frac{3}{4}$ of a dollar a day, how long will it take a man to earn $\$9$?

3. How many chairs, at $\$5\frac{1}{2}$ apiece, can be bought for $\$38\frac{1}{2}$?

4. How many bushels of wheat, at $\$2\frac{3}{8}$ a bushel, can be bought for $\$19$?

5. If a man can build $4\frac{2}{5}$ rods of wall in a day, how many days will it take him to build $13\frac{1}{5}$ rods?

6. If 9 pounds of tea cost $\$5\frac{5}{8}$, what does it cost a pound?

7

7. How many pounds of sugar can be bought for 50 cents, when it is $8\frac{1}{2}$ cents a pound?

8. If 7 bushels of potatoes cost $\$8\frac{2}{5}$, how much are they a bushel?

9. If you spend $\frac{3}{4}$ of a dollar a week, how long will you be in spending $\$6$?

10. How many times $\frac{2}{3}$ in 4? In 6?

11. How many times $2\frac{1}{3}$ in 12? In 9?

12. How many times 3 in $\frac{18}{5}$? In $2\frac{4}{5}$?

13. How many times 8 in $9\frac{1}{4}$? In $10\frac{1}{5}$?

14. How many times 4 in $11\frac{1}{4}$? In $12\frac{1}{4}$?

15. How many times $\frac{2}{3}$ are $13\frac{1}{2}$?

16. How many times $\frac{5}{6}$ are $12\frac{3}{4}$?

17. How many times $\frac{6}{7}$ are $5\frac{4}{7}$?

18. How many times $1\frac{1}{2}$ are $6\frac{1}{4}$?

19. How many times $1\frac{1}{3}$ are $6\frac{2}{3}$?

20. How many times $2\frac{1}{6}$ are $8\frac{1}{6}$?

21. How many times $6\frac{1}{6}$ are $12\frac{3}{6}$?

22. How many times $5\frac{1}{7}$ are 2 times $4\frac{3}{7}$?

23. How many times $5\frac{1}{6}$ are 3 times $6\frac{2}{5}$?

24. How many times 7 are 6 times $4\frac{4}{6}$?

25. How many times $10\frac{1}{2}$ are 7 times $9\frac{1}{2}$?

26. How many times $33\frac{1}{3}$ are 6 times $16\frac{2}{3}$?

27. How many times 3 times $2\frac{1}{5}$ are $10\frac{2}{5}$?

28. When eggs are $\frac{1}{6}$ of a dollar a dozen, how many dozen must be given for 16 pounds of coffee, at $\frac{1}{8}$ of a dollar a pound?

29. If 3 pounds of butter cost $\frac{3}{4}$ of a dollar, what cost 10 pounds?

30. If a teacher distributed in rewards among his pupils $\$7\frac{1}{2}$, by giving $\frac{1}{4}$ of a dollar to each, how many pupils did he reward?

31. How many lambs, at $\$2\frac{1}{4}$ apiece, may be had for 10 calves, at $\$6\frac{3}{4}$ apiece?

32. How many bushels of apples, at $\frac{5}{16}$ of a dollar a bushel, will pay for 15 yards of calico, at $\frac{3}{8}$ of a dollar a yard?

33. If 7 yards of broadcloth cost 37\frac{8}{10}$, what will 8 yards cost?

34. If Joseph can walk 6$\frac{2}{7}$ miles in 2 hours, how many miles can he walk in 8 hours?

35. If one man consumes 1$\frac{3}{8}$ pounds of meat in one day, how many men will consume 15$\frac{1}{8}$ pounds in the same time?

36. If 4 men can do a piece of work in 16$\frac{1}{3}$ days, how long will it take 3 men to do the same?

37. How many bushels of potatoes, at $\frac{3}{16}$ of a dollar a peck, must be given for 10 bushels of wheat, at 2\frac{4}{16}$ a bushel?

38. If a man can mow 1$\frac{1}{5}$ acres in one day, how long will it take 6 men to mow 14$\frac{2}{5}$ acres?

LESSON XXXIII.

1. What is $\frac{1}{3}$ of 2?

NOTE. $\frac{1}{3}$ of 1 is $\frac{1}{3}$, and $\frac{1}{3}$ of 2 is 2 times $\frac{1}{3}$, which are $\frac{2}{3}$.

2. What is $\frac{1}{3}$ of 5? Of 7? Of 8? Of 11? Of 13? Of 17? Of 18?

3. What is $\frac{1}{4}$ of 2? Of 3? Of 7? Of 9? Of 10? Of 13? Of 14?

4. What is $\frac{1}{5}$ of 3? Of 5? Of 8? Of 13? Of 17? Of 19? Of 20?

5. What is $\frac{1}{7}$ of 2? Of 3? Of 5? Of 11? Of 12? Of 17? Of 18?

6. What is $\frac{1}{8}$ of 4? Of 5? Of 7? Of 12? Of 18? Of 21? Of 23?

7. What is $\frac{1}{10}$ of 3? Of 5? Of 6? Of 7? Of 11? Of 19? Of 20?

8. What is $\frac{1}{6}$ of 4? Of 7? Of 9? Of 10? Of 13? Of 15? Of 16?

9. What is $\frac{2}{3}$ of 7? Of 8? Of 9? Of 10? Of 11? Of 12? Of 14?

10. What is $\frac{3}{4}$ of 2? Of 7? Of 9? Of 13? Of 14? Of 18? Of 19?

11. How many are $\frac{2}{5}$ of 3? Of 5? Of 7? Of 9? Of 17? Of 20?

12. How many are $\frac{5}{6}$ of 2? Of 3? Of 9? Of 15? Of 20? Of 25?

13. How many are $\frac{3}{7}$ of 5? Of 6? Of 10? Of 11? Of 12? Of 31?

14. How many are $\frac{4}{9}$ of 2? Of 3? Of 6? Of 10? Of 13? Of 20?

15. How many are $\frac{5}{11}$ of 2? Of 4? Of 7? Of 9? Of 10? Of 12?

16. If 4 oranges are divided equally among 5 boys, how much of an orange will each boy receive?

17. If 8 bushels of apples cost \$5, what part of a dollar costs 1 bushel? 2 bushels?

18. If 12 chairs cost \$9, what costs 1 chair?

19. If 8 bushels of corn cost \$6, what cost 3 bushels?

20. What cost $\frac{2}{5}$ of a pound of butter, if 1 pound costs 17 cents?

21. At 19 cents a yard, what cost $\frac{3}{4}$ of a yard of calico?

22. A hogshead of molasses was sold for \$33, which was $\frac{7}{6}$ of the cost; what was the cost?

23. Paid \$101 dollars for a horse, which was $\frac{4}{5}$ of his real value; what was his real value?

24. After spending $\frac{11}{13}$ of my money, I had \$20 left; how much had I at first?

25. A gentleman is 83 years old, and his son is $\frac{3}{5}$ as old; how old is his son?

26. What cost 2 dozen eggs, at the rate of 3 for 7 cents?

27. What cost 9 bushels of quinces, at the rate of 4 bushels for \$7?

28. If 5 oranges cost 28 cents, what will 11 oranges cost?

29. If $31 is $\frac{3}{7}$ of what money William has, how much has he?

30. A boy having 66 cents, gave $\frac{3}{11}$ of them to his mother; how many did he give her?

31. Sold a horse for $64, which was $\frac{4}{9}$ of what he cost me; what did he cost me?

32. $\frac{1}{4}$ of Edward's money is 5 times Henry's, and Henry has $11; how much has Edward?

33. $\frac{2}{3}$ of 24 is $\frac{4}{15}$ as many books as James owns; how many does he own?

34. $\frac{4}{7}$ of 28 is $\frac{2}{3}$ of what number?

35. $\frac{1}{8}$ of 16 is $\frac{1}{4}$ of what number?

36. $\frac{5}{6}$ of 42 is $\frac{2}{3}$ of what number?

37. $\frac{3}{4}$ of 48 is $\frac{6}{5}$ of what number?

38. $\frac{4}{11}$ of 22 is $\frac{8}{10}$ of what number?

39. $\frac{7}{8}$ of 32 is $\frac{2}{5}$ of what number?

40. $\frac{6}{8}$ of 56 is $1\frac{1}{4}$ times what number?

41. $\frac{4}{5}$ of 75 is $2\frac{2}{3}$ times what number?

• 42. Bought a cow and a horse; the cost of the cow was $32, and that of the horse was 2 times $\frac{5}{8}$ of the cost of the cow; what was the cost of the horse?

43. If 3 quarts of cranberries, at 80 cents a peck, will pay for 4 pounds of rice, how much is the rice a pound?

44. $\frac{5}{6}$ of 48 is how many times $\frac{1}{7}$ of 35?

45. $\frac{7}{8}$ of 72 is how many times $\frac{3}{4}$ of 12?

LESSON XXXIV.

1. William gave Edward $\frac{1}{2}$ of an apple; how many fourths did he give him?

NOTE.—Since $1 = \frac{4}{4}$, $\frac{1}{2}$ of $1 = \frac{2}{4}$; or, since $1 = \frac{2}{2}$, and $1 = \frac{4}{4}$, there are twice as many fourths as halves in a number.

2. John had $\frac{1}{3}$ of a dollar; how many sixths had he?

3. A man bought $\frac{1}{3}$ of an acre of land at one time, and $\frac{2}{3}$ of an acre at another time; how many thirds did he buy?

4. If you wish to give $\frac{1}{2}$ of a bushel of corn to one man, and $\frac{1}{3}$ to another, how many sixths would you give them both?

5. Gave $\frac{1}{4}$ of a dollar to Lydia, and $\frac{2}{4}$ to Sarah; to which was the most given?

6. Sold $\frac{1}{2}$ of an acre of land to one man, $\frac{2}{4}$ to another, and $\frac{1}{4}$ to another; how many fourths were sold? How many acres?

7. $\frac{1}{2} + \frac{3}{4} + \frac{1}{4}$ are how many fourths? How many times 1?

8. A man had a quantity of fruit, which he wished to divide among 3 of his friends; he gave $\frac{1}{2}$ to one, $\frac{1}{4}$ to another, and $\frac{3}{8}$ to another; how many eighths did each receive?

9. $\frac{1}{2} + \frac{1}{4} + \frac{3}{8}$ are how many eighths? How many times 1?

10. A gentleman kept 3 fires in his house, during the winter; the first fire consumed $\frac{2}{3}$ of a ton of coal, the second $\frac{1}{2}$, and the third $\frac{1}{6}$ of a ton; how many sixths did each consume? How many tons did they all consume?

11. $\frac{2}{3} + \frac{1}{2} + \frac{1}{6}$ are how many sixths? How many times 1?

12. $\frac{1}{2}$ is how many tenths? How many twentieths?

13. $\frac{1}{4}$ is how many eighths? How many sixteenths?

14. $\frac{2}{3}$ is how many thirds? How many sixths?

15. $\frac{3}{8}$ is how many fourths? How many sixteenths?

16. $\frac{1}{3}$ is how many ninths? How many twelfths?

17. $\frac{3}{18}$ is how many sixths? How many twelfths?

18. Reduce $\frac{1}{2}$ and $\frac{2}{3}$ to a common denominator.

NOTE. — Fractions are said to have a common denominator when their denominators are alike. *Multiplying both the numerator and denominator by the same number, does not alter the value of a fraction.*

19. Reduce $\frac{1}{4}$ and $\frac{2}{5}$ to a common denominator.

20. Reduce $\frac{1}{4}$, $\frac{4}{5}$, and $\frac{5}{8}$ to a common denominator.

21. Change $\frac{3}{4}$ and $\frac{3}{12}$ to a common denominator.

22. Change $\frac{3}{8}$ and $\frac{3}{12}$ to fourths.

23. Change $\frac{1}{4}$, $\frac{3}{4}$, and $\frac{5}{8}$ to a common denominator.

24. A farmer sold $\frac{3}{4}$ of a bushel of peaches to one man, $\frac{2}{5}$ to another, and $\frac{2}{10}$ to another; how many bushels did he sell in all?

25. What is the sum of $\frac{5}{7}$ and $\frac{4}{5}$?

26. What is the sum of $\frac{3}{4}$ and $\frac{5}{6}$?

27. What is the sum of $\frac{4}{4}$ and $\frac{5}{7}$?

28. What is the sum of $\frac{7}{8}$ and $\frac{4}{9}$?

29. $\frac{5}{7} + \frac{7}{8}$ are how many times 1?

30. $\frac{3}{11} + \frac{4}{3}$ are how many times 1?

31. $1\frac{5}{7} + 2\frac{1}{2}$ are how many times 1?

32. $11\frac{1}{4} + 10\frac{1}{2}$ are how many times 1?

33. What is the sum of $\frac{1}{2} + \frac{1}{6} + \frac{2}{5}$?

34. What is the sum of $\frac{4}{5} + \frac{3}{10} + \frac{3}{4}$?

35. What is the sum of $3\frac{4}{9} + \frac{1}{3} + \frac{4}{6}$?

36. What is the sum of $11 + \frac{1}{4} + \frac{7}{8}$?

37. $\frac{3}{4}$ from $\frac{7}{8}$ leaves how many?

38. $\frac{1}{4}$ from $\frac{3}{4}$ leaves how many?

39. $\frac{6}{11}$ from $\frac{19}{12}$ leaves how many?

40. $\frac{2}{3}$ from 12 leaves how many?

41. $\frac{5}{8}$ from $\frac{17}{4}$ leaves how many?

42. $10\frac{1}{4}$ less $1\frac{3}{8}$ are how many?

43. $\frac{19}{2}$ less $\frac{5}{8}$ are how many?

44. $11\frac{3}{4} - 10\frac{1}{4}$ leaves how many?

45. $16 - 1\frac{7}{8}$ leaves how many?

46. $\frac{1}{3} + \frac{1}{4}$ are how much less than a whole one?

47. $\frac{1}{8} + \frac{1}{2}$ are how much less than a whole one?

48. $\frac{1}{5} + \frac{2}{3} + \frac{3}{4}$ are how much less than 2 ?

49. When have fractions a common denominator ?

50. How do you reduce fractions to a common denominator ?

51. Does multiplying or dividing both the numerator and denominator by the same number alter the value of a fraction ?

52. What number added to $\frac{2}{3}$ of 31 will make 25 ?

53. What number taken from $\frac{3}{4}$ of 26 will leave $16\frac{1}{2}$?

54. If you should sell $\frac{1}{6}$ of your apples, and give away $\frac{1}{2}$, what part would you have remaining ?

55. George, being asked his age, replied, if he were $\frac{1}{3}$ and $\frac{1}{6}$ older he should be 40 years old; how old was he ?

56. Joseph deposited $\frac{1}{4}$ of his money in the savings-bank, paid $\frac{2}{3}$ for tuition, and then found $\frac{1}{2}$ of the remainder was just $9; how much money had he in all ?

LESSON XXXV.

1. John had $\frac{1}{4}$ of a dollar, and gave $\frac{1}{2}$ of it to James; what part of a dollar did James receive ?

2. William had $\frac{1}{4}$ of a melon, and gave his brother $\frac{1}{3}$ of it; what part of a whole melon did he give away ?

3. What is $\frac{1}{2}$ of $\frac{1}{5}$? $\frac{1}{3}$ of $\frac{1}{4}$?

4. A gentleman owned $\frac{1}{6}$ of a ship; if he should sell $\frac{1}{4}$ of his share, what part of the ship would he sell ?

5. What is $\frac{1}{4}$ of $\frac{1}{6}$? $\frac{1}{6}$ of $\frac{1}{4}$?

6. Mary gathered $\frac{1}{2}$ of a box of strawberries, and eat $\frac{1}{4}$ of them; what part of a box did she eat ?

7. What is $\frac{1}{3}$ of $\frac{1}{2}$? $\frac{1}{4}$ of $\frac{1}{3}$? $\frac{1}{2}$ of $\frac{1}{4}$? $\frac{1}{4}$ of $\frac{1}{3}$?

8. What is $\frac{1}{2}$ of $\frac{1}{4}$? $\frac{1}{4}$ of $\frac{1}{4}$? $\frac{1}{5}$ of $\frac{1}{4}$? $\frac{1}{7}$ of $\frac{1}{4}$?

9. What is $\frac{1}{3}$ of $\frac{1}{3}$? $\frac{1}{5}$ of $\frac{1}{5}$? $\frac{1}{6}$ of $\frac{1}{6}$? $\frac{1}{7}$ of $\frac{1}{7}$?

10. A gentleman, owning $\frac{2}{5}$ of a farm, sold $\frac{1}{3}$ of his part; what part of the whole farm did he sell?

11. What part of an acre is $\frac{4}{5}$ of $\frac{7}{11}$ of an acre?

NOTE. $\frac{1}{5}$ of $\frac{7}{11} = \frac{7}{55}$; and $\frac{4}{5}$ of $\frac{7}{11} = 4$ times $\frac{7}{55} = \frac{28}{55}$.

12. What part of a ship is $\frac{3}{4}$ of $\frac{2}{3}$ of it?

13. If you should have $\frac{5}{7}$ of a barrel of apples, and sell $\frac{4}{5}$ of them, what part of a barrel would you sell?

14. What is $\frac{1}{2}$ of $\frac{2}{3}$? $\frac{2}{3}$ of $\frac{3}{7}$? $\frac{3}{7}$ of $\frac{7}{9}$? $\frac{3}{8}$ of $\frac{5}{6}$?

15. What is $\frac{4}{9}$ of $\frac{3}{7}$? $\frac{3}{4}$ of $\frac{5}{8}$? $\frac{5}{7}$ of $\frac{9}{10}$? $\frac{2}{9}$ of $\frac{3}{4}$? $\frac{4}{10}$ of $\frac{5}{6}$?

16. What is $\frac{7}{11}$ of $\frac{1}{5}$? $\frac{5}{8}$ of $\frac{2}{7}$? $\frac{2}{3}$ of $1\frac{1}{2}$? $\frac{4}{9}$ of $1\frac{9}{11}$? $\frac{3}{5}$ of $\frac{1}{7}$?

17. What is $\frac{3}{8}$ of $\frac{4}{5}$? $\frac{3}{4}$ of $\frac{9}{10}$? $\frac{7}{8}$ of $\frac{4}{10}$? $\frac{9}{10}$ of $1\frac{1}{3}$? $\frac{4}{7}$ of $\frac{7}{12}$?

18. If a barrel of potatoes cost $2\frac{1}{2}$, what will $\frac{1}{3}$ of a barrel cost?

NOTE. $2\frac{1}{2} = \frac{5}{2}$, and $\frac{1}{3}$ of $\frac{5}{2} = \frac{5}{6}$.

19. If 4 pounds of sugar cost $33\frac{1}{3}$ cents, what cost 1 pound? What is $\frac{1}{4}$ of $33\frac{1}{3}$?

20. Bought 6 yards of cloth for $12\frac{3}{4}$; what was the cost of one yard? What is $\frac{1}{6}$ of $12\frac{3}{4}$?

21. Sold 9 barrels of apples for $18\frac{3}{4}$; what cost 1 barrel? What is $\frac{1}{9}$ of $18\frac{3}{4}$?

22. What is $\frac{1}{3}$ of $4\frac{1}{2}$? $\frac{2}{3}$ of $3\frac{1}{4}$? $\frac{1}{4}$ of $5\frac{1}{4}$? $\frac{4}{5}$ of $10\frac{1}{2}$?

23. What is $\frac{5}{7}$ of $11\frac{1}{3}$? $\frac{1}{8}$ of $7\frac{1}{5}$? $\frac{7}{10}$ of $16\frac{3}{4}$? $\frac{2}{3}$ of $11\frac{1}{6}$?

24. Gave $\frac{2}{3}$ of $4\frac{1}{4}$ for a hat; what did the hat cost?

25. If a pine-apple cost $9\frac{3}{8}$ cents, what will $\frac{3}{4}$ of one cost?

26. John can reap an acre in $12\frac{1}{2}$ hours, and George as much in $\frac{3}{4}$ of that time; how long will it take George to reap an acre?

27. What cost $\frac{4}{5}$ of a bushel of corn, at $\frac{7}{8}$ of a dollar a bushel?

28. If a bushel of apples will buy $\frac{5}{8}$ of a bushel of peaches, what quantity of peaches can be bought for $\frac{4}{5}$ of a bushel of apples?

29. What cost 3 quarts of beans, if 1 peck cost $\frac{7}{8}$ of a dollar?

30. If 6 pounds of sugar cost $\frac{5}{8}$ of a dollar, what cost 5 pounds?

31. If 3 tons of hay cost $\$31\frac{1}{2}$, what cost $\frac{3}{4}$ of a ton?

32. If a man earn $\$8\frac{4}{5}$ in 6 days, how much can he earn in $\frac{1}{4}$ of $\frac{4}{5}$ of a day?

33. When $\frac{3}{4}$ of $\$6\frac{1}{4}$ will buy a yard of broadcloth, how much will $\frac{3}{11}$ of a yard cost?

34. Sold $\frac{4}{5}$ of $5\frac{1}{2}$ pounds of coffee, at $1\frac{1}{2}$ dimes a pound; how much did it come to?

35. Andrew is 20 years old, and his age exceeds by 5 years $\frac{3}{4}$ of $\frac{4}{5}$ of Walter's age; how old is Walter?

LESSON XXXVI.

1. How many yards of cloth, at $\frac{1}{4}$ of a dollar a yard, can be bought for $\$2$? How many times $\frac{1}{4}$ in 2?

NOTE. $2 = \frac{8}{4}$; and $\frac{1}{4}$ is contained in $\frac{8}{4}$ 8 times.

2. Susan distributed 4 pears among some schoolmates, giving to each $\frac{2}{3}$ of a pear; how many schoolmates were there?

3. If you spend $\frac{3}{4}$ of a dollar a week, how long will $\$6$ last you?

4. How many times $\frac{3}{4}$ in 6? $\frac{4}{5}$ in 9?

5. If $1\frac{1}{3}$ will pay 1 man for a day's work, how many men will $6\frac{2}{3}$ pay?

6. How many pens, at $1\frac{1}{2}$ cents apiece, can be bought for 12 cents?

7. How many times $2\frac{1}{4}$ in 12? $2\frac{3}{4}$ in $9\frac{1}{8}$?

8. At $2\frac{1}{4}$ a pair, how many pairs of shoes can be bought for $15\frac{3}{4}$?

9. If $3\frac{2}{3}$ pounds of butter last a family 1 week, how long will $25\frac{2}{3}$ pounds last the same family?

10. How many times is $3\frac{1}{5}$ contained in $12\frac{1}{5}$?

11. How many times is $1\frac{3}{7}$ contained in $11\frac{3}{7}$?

12. How many times is $\frac{2}{3}$ contained in $\frac{5}{6}$?

NOTE. $\frac{2}{3} = \frac{4}{6}$; and $\frac{4}{6}$ is contained in $\frac{5}{6}$ $1\frac{1}{4}$ times; or, 1 is contained in $\frac{5}{6}$ $\frac{5}{6}$ times; and $\frac{1}{3}$ is contained in $\frac{5}{6}$ 8 times $\frac{5}{6}$ $= \frac{15}{6}$ times; and $\frac{2}{3}$ is contained in $\frac{5}{6}$ $\frac{1}{2}$ of $\frac{15}{6}$ times $= 1\frac{1}{4}$ times.

13. How many times is $\frac{2}{3}$ contained in $\frac{17}{8}$?

14. If a horse eat $\frac{1}{6}$ of a ton of hay in 1 week, how many weeks will he be in eating $\frac{8}{9}$ of a ton?

15. How many times is $\frac{3}{4}$ contained in $\frac{9}{10}$?

16. How many pounds of tea, at $\frac{2}{3}$ of a dollar a pound, can be bought for $\frac{9}{12}$ of a dollar?

17. How many times is $\frac{4}{5}$ contained in $2\frac{1}{3}$?

18. How many times is $\frac{4}{7}$ contained in $3\frac{1}{2}$?

19. How many times is $3\frac{1}{5}$ contained in $8\frac{1}{8}$?

20. How many times is $\frac{4}{9}$ contained in $11\frac{1}{2}$?

21. How many times is $5\frac{3}{8}$ contained in $12\frac{1}{4}$?

22. How many times is $3\frac{1}{6}$ contained in $7\frac{1}{12}$?

23. How many times is $6\frac{1}{4}$ contained in $10\frac{5}{8}$?

24. If a lamb cost $2\frac{3}{5}$, how many can be bought for $10\frac{2}{5}$?

25. At $\frac{2}{5}$ of a dollar a yard, how many yards of cloth can be bought for $4\frac{1}{2}$?

26. If 6 pounds of cheese cost $\frac{4}{7}$ of a dollar, what does 1 pound cost?

27. How many bushels of apples, at $\frac{2}{3}$ of a dol-

lar a bushel, must be given for 7 bushels of potatoes, at $\frac{3}{4}$ of a dollar a bushel?

28. If a man can walk $\frac{2}{3}$ of a mile in $\frac{1}{4}$ of an
hour, in how many hours can he walk $8\frac{1}{3}$ miles?

29. How many pounds of sugar, at $\frac{1}{12}$ of a dollar a pound, can be bought for $\frac{3}{4}$ of a bushel of
cranberries, at $ $3\frac{1}{2}$ a bushel?

30. What will $1\frac{1}{4}$ yards of cloth cost, if $\frac{3}{5}$ of a
yard cost $\frac{3}{8}$ of a dollar?

31. A farmer, having sold a load-of wood for
$ $7\frac{1}{2}$, spent $\frac{2}{5}$ of the money for tea, at $\frac{3}{4}$ of a dollar
a pound, and the balance for coffee, at $\frac{1}{6}$ of a dollar
a pound; how many pounds of each did he obtain?

32. If $\frac{2}{3}$ of a barrel of beef cost $ $3\frac{3}{4}$, what will
$\frac{3}{8}$ of a barrel cost?

33. How many times is $\frac{1}{5}$ of $\frac{3}{4}$ contained in $\frac{4}{7}$ of
$2\frac{1}{5}$?

34. When oats are $\frac{5}{16}$ of a dollar a bushel, and
corn $\frac{7}{8}$ of a dollar a bushel, how many bushels of
oats must be given for 5 bushels of corn?

LESSON XXXVII.

1. What part of 3 is 1? Of 3 is 2?

NOTE. 1 is $\frac{1}{3}$ of 3, since 1 taken 3 times $= 3$; and since 1
is $\frac{1}{3}$ of 3, $2 = \frac{2}{3}$ of 3.

2. What part of 5 is 1? Of 5 is 3?
3. What part of 9 is 1? Of 9 is 5?
4. What part of 7 is 3? Of 8 is 5?
5. What part of 11 is 5? Of 13 is 6?
6. What part of 6 is 7? Of 9 is 11?
7. What part of 13 is 14? Of 14 is 17?
8. What part of 2 is $\frac{1}{4}$? Of $\frac{3}{4}$ is $\frac{8}{5}$?

NOTE. $2 = \frac{8}{4}$, and $\frac{1}{4}$ is the same part of $\frac{8}{4}$ that 1 is of 8,
which is $\frac{1}{8}$; and $\frac{1}{4} = \frac{15}{20}$, $\frac{8}{5} = \frac{8}{20}$, and $\frac{8}{20}$ is the same part of
$\frac{15}{20}$ that 8 is of 15, which is $\frac{8}{15}$.

9. What part of 5 is $\frac{1}{3}$? Of 7 is $\frac{4}{5}$?
10. What part of 9 is $\frac{1}{4}$? Of 10 is $\frac{5}{6}$?
11. What part of $\frac{7}{8}$ is 3? Of $\frac{5}{6}$ is 5?
12. What part of $\frac{9}{7}$ is $\frac{1}{3}$? Of $\frac{5}{9}$ is $\frac{1}{5}$?
13. What part of $\frac{9}{10}$ is $\frac{1}{6}$? Of $\frac{8}{11}$ is $\frac{1}{4}$?
14. What part of $\frac{8}{9}$ is $\frac{3}{4}$? Of $\frac{5}{7}$ is $\frac{2}{3}$?
15. What part of $\frac{1}{4}$ is $\frac{4}{5}$? Of $\frac{2}{3}$ is $\frac{6}{11}$?
16. What part of 1 peck is 5 quarts?
17. What part of 1 pound is 9 shillings?
18. What part of 1 quarter is $14\frac{1}{2}$ pounds?
19. What part of 2 dimes is 13 cents?
20. What part of 1 week 4 days is 5 days?
21. What part of 5 bushels is 3 pecks 6 quarts?
22. What part of 10 pounds is 4 pounds 12 ounces?
23. What part of 2 hogsheads is 1 hogshead 17 gallons?
24. What part of 3 yards 3 quarters is 3 quarters 3 nails?
25. What part of 3 leagues 2 miles is 1 mile 3 furlongs?
26. What part of 1 acre 2 roods is 3 roods 15 square rods?
27. What part of a mile is 3 furlongs 15 rods?
28. What part of 8 yards is 5 ells English 3 quarters?
29. What part of a pint is $\frac{1}{12}$ of a gallon?

NOTE. — Since 1 gallon = 4 quarts, $\frac{1}{12}$ of a gallon = $\frac{1}{12}$ of 4 quarts = $\frac{4}{12}$ of a quart; and since 1 quart = 2 pints, $\frac{4}{12}$ of a quart = $\frac{4}{12}$ of 2 pints = $\frac{8}{12}$ of a pint = $\frac{2}{3}$ of a pint.

30. What part of an inch is $\frac{1}{48}$ of a yard?
31. What part of a pennyweight is $\frac{2}{8}$ of an ounce?
32. What part of a rod is $\frac{1}{400}$ of a mile?
33. What part of a day is $\frac{2}{15}$ of a week?
34. What part of a gill is $\frac{2}{65}$ of a gallon?

35.   What part of a shilling is $\frac{4}{5}$ of a farthing ?

NOTE. — Since 4 farthings = 1 penny, 1 farthing = $\frac{1}{4}$ of a
penny, and $\frac{4}{5}$ of a farthing = $\frac{4}{5}$ of $\frac{1}{4}$ of a penny = $\frac{4}{20}$ of a
penny ; and since 12 pence = 1 shilling, 1 penny = $\frac{1}{12}$ of a
shilling, and $\frac{4}{20}$ of a penny = $\frac{4}{20}$ of $\frac{1}{12}$ of a shilling = $\frac{4}{240}$,
or $\frac{1}{60}$ of a shilling.

36.   What part of a week is $\frac{2}{3}$ of a day ?
37.   What part of a yard is $\frac{3}{10}$ of an inch ?
38.   What part of a ton is $\frac{1}{4}$ of a pound ?
39.   What part of a bushel is $\frac{2}{3}$ of a pint ?
40.   What part of a pound is $\frac{1}{2}$ of a pennyweight ?
41.   What is the value of $\frac{1}{8}$ of a hogshead in
gallons ?
42.   What is the value of $\frac{3}{8}$ of a pound in shillings and pence ?

NOTE. — Since 1 pound = 20 shillings, $\frac{3}{8}$ of a pound = $\frac{3}{8}$ of
20 shillings = $7\frac{4}{8}$ shillings = $7\frac{1}{2}$ shillings ; and since 1 shilling
= 12 pence, $\frac{1}{2}$ of a shilling = $\frac{1}{2}$ of 12 pence = 6 pence ; therefore, $\frac{3}{8}$ of a pound = 7 shillings 6 pence.

43.   What is the value of $\frac{3}{8}$ of a pound Troy ?
44.   What is the value of $\frac{5}{8}$ of a yard ?
45.   What is the value of $\frac{3}{4}$ of a hogshead ?
46.   What is the value of $\frac{4}{5}$ of a bushel ?
47.   What is the value of $\frac{3}{16}$ of a ton weight ?
48.   What is the value of $\frac{3}{11}$ of a week ?

LESSON XXXVIII.

1.   A market-woman sold some butter, eggs and
milk ; for the eggs and milk she received $4, which
was $\frac{2}{3}$ of what she received for the butter ; how
much did she receive for the butter ?
2.   Sold $\frac{3}{4}$ of an acre of land for $24 ; at how
much an acre was it sold ?
3.   If James can run 60 rods in a minute, and

John $\frac{2}{5}$ as far, how long will it take John to run $15\frac{1}{4}$ rods?

4. 28 is $\frac{7}{8}$ of a number; what is that number?

5. 36 is $\frac{6}{11}$ of a number; what is that number?

6. If $\frac{4}{5}$ of a firkin of butter cost $8, what will $\frac{1}{4}$ of a firkin cost?

7. $\frac{1}{3}$ of $\frac{1}{4}$ of 24 is $\frac{2}{3}$ of what number?

8. $\frac{1}{3}$ of $\frac{2}{5}$ of 15 is $\frac{3}{7}$ of what number?

9. If a bushel of potatoes is $\frac{2}{5}$ of a barrel, and cost 1\frac{1}{4}$, how much will 1 barrel cost?

10. A and B own some wood together; A's share is $\frac{4}{9}$; it is sold, and A receives as $\frac{1}{2}$ of his part of the proceeds $24; how much did the wood sell for? How much was B's part of the proceeds?

11. A teacher, being asked how many scholars he had, answered that 20 of them were girls, and $\frac{2}{3}$ of them and 4 more were boys; how many boys were there? How many scholars in all?

12. 14 is $\frac{2}{3}$ of $\frac{7}{8}$ of what number?

13. 18 is $\frac{3}{4}$ of $\frac{3}{5}$ of what number?

14. John Jones sold a horse for $60, which was 2 times $\frac{3}{5}$ of what he gave for it; how much did he gain by the sale?

15. Bought a piece of land for $64, and sold $\frac{1}{2}$ of $\frac{4}{5}$ of it for $28; how much will be gained if the rest be sold at the same rate?

16. $\frac{4}{5}$ of a pole is above ground, and 3 feet is $\frac{1}{4}$ of the part in the ground; what is the length of the pole?

17. A gentleman paid away a sum of money: $\frac{3}{7}$ to James Brown, $\frac{2}{3}$ of $\frac{5}{8}$ of the remainder, which was $15, to Peter Smith, and the balance to Edward Robinson; what was the whole amount, and how much did Brown and Robinson each receive?

18. 35 is $\frac{5}{8}$ of $\frac{7}{8}$ of what number?

19. When 2\frac{1}{4}$ will purchase $\frac{3}{4}$ of a barrel of

flour, what part of a barrel can be purchased for $\$4\frac{1}{2}$?

20. If $\frac{2}{3}$ of a dozen of eggs cost $\frac{1}{8}$ of a dollar, how many dozen can be bought for $\$1$?

21. 36 is $\frac{3}{4}$ of how many times 8?

22. 52 is $\frac{4}{5}$ of how many times 13?

23. John Doe and Richard Roe enter into a speculation together, with a capital of $\$160$, $\frac{5}{6}$ of $\frac{3}{8}$ of which was contributed by John, and the remainder by Richard; on dividing their gains, John received $\$60$ as his share; how much was Richard's share?

24. Henry spent $\frac{1}{4}$ of his money for pencils, $\frac{2}{3}$ for a slate, and had 4 cents left; how much money had he, and how many pencils did he buy, at 2 cents apiece?

25. 30 is $\frac{5}{6}$ of how many times 9?

26. 64 is $\frac{8}{9}$ of how many times 12?

27. If $\frac{2}{5}$ of Robert's age has passed since he was 21 years old, how old is he?

28. $\frac{4}{7}$ of $\frac{5}{8}$ is how many times $\frac{1}{7}$?

29. Susan is 25 years old, and Lydia is only $6\frac{1}{4}$ years; how many times is Susan as old as Lydia?

30. A man sold $\frac{3}{7}$ of a barrel of cider, and $\frac{1}{4}$ of what was left was worth $\$1\frac{1}{5}$; what was the whole barrel worth?

31. What cost $\frac{1}{9}$ of a hogshead of molasses, at $\frac{3}{5}$ of a dollar a gallon?

32. At $\frac{1}{12}$ of a dollar a quart, what part of a bushel of walnuts can be bought for $\$2\frac{1}{3}$?

33. Allowing 10 working hours to a day, in what part of a week, consisting of 6 working days, can a man earn $\$7$, if his pay be at the rate of $\frac{1}{5}$ of a dollar an hour?

34. If $1\frac{1}{4}$ pecks be added to $\frac{3}{16}$ of a bushel, what part of a bushel will there then be?

35. If from $\frac{5}{8}$ of a hundred-weight there be

taken $\frac{1}{4}$ of a quarter, what part of a hundred-weight will then remain ?

36. A young man lost $\frac{1}{4}$ of his capital in trade, but afterwards gained $100, which made his capital $1000 ; how much money did he lose?

37. 60 is $\frac{6}{7}$ of how many times $\frac{1}{8}$ of 30 ?

38. 84 is $\frac{1}{7}$ of how many times $\frac{1}{2}$ of 22 ?

39. A watch-chain cost $48, and $\frac{2}{3}$ of the cost of the chain was $\frac{2}{5}$ of the cost of the watch ; what was the cost of the watch?

LESSON XXXIX.

1. If 7 is $\frac{2}{3}$ of some number, what is $\frac{1}{7}$ of the same number ?

2. If 10 is $\frac{1}{4}$ of some number, what is $\frac{2}{5}$ of the same number ?

3. If 12 is $\frac{1}{6}$ of some number, what is $\frac{3}{8}$ of the same number ?

4. What number added to 2 times $\frac{2}{3}$ of 37 will make the number 18 ?

5. What number added to 5 times $\frac{3}{7}$ of 30 will make the number 65 ?

6. What number taken from 4 times $\frac{5}{6}$ of 19 will leave $50\frac{1}{4}$?

7. What number taken from 3 times $\frac{5}{3}$ of 22 will leave 100 ?

8. What number is that to which if $\frac{2}{5}$ of itself be added, the number will be 48 ?

9. What number is that to which if its half and fourth be added, the sum will be 100 ?

10. A lady, being asked how old she was, answered, that if $\frac{1}{3}$ and $\frac{1}{6}$ of her age were added to her age, it would be 99 years ; how old was she ?

11. Paid $\frac{1}{5}$ of my money for pencils, $\frac{1}{3}$ for an account-book, and 16 cents for india-rubber, and

had $\frac{1}{5}$ as much left as I had in the beginning; how much had I at first?

12. A father left his son a legacy, $\frac{1}{4}$ of which he spent in 6 months, and $\frac{3}{4}$ of the remainder lasted him 8 months longer, when he had only $100 remaining; what sum did his father leave him?

13. A and B made an even exchange of horses; by the trade A lost 24 dollars, which was $\frac{2}{7}$ of the value of his horse; what was the value of each horse?

14. If one man can cut $1\frac{1}{2}$ cords of wood in a day, how long will it take 3 men to cut the same?

15. If 3 horses consume $\frac{3}{4}$ of a bushel of oats in 2 days, how many horses will consume $3\frac{1}{2}$ bushels in the same time? ·

16. $\frac{2}{3}$ of 36 is $\frac{4}{5}$ of how many tenths of 20?

17. $\frac{3}{4}$ of 44 is $\frac{3}{5}$ of how many sevenths of 35?

18. $\frac{4}{5}$ of 30 is $\frac{3}{8}$ of how many ninths of 45?

19. How many yards of cloth that is $\frac{3}{4}$ of a yard wide, are equal to 9 yards that is $\frac{1}{2}$ of a yard wide?

20. Simon bought 2 dozen of eggs, at 6 cents a dozen, 2 pounds of beef, at 12 cents, and spent $\frac{1}{5}$ of what he had in the beginning for vegetables, and $\frac{1}{4}$ for a lobster, and had $\frac{11}{35}$ left; how much had he at first?

21. If a man, by selling a cart for $1\frac{2}{3}$ times its cost, gain $11, what was the cost of the cart?

22. Robert receives $\frac{3}{4}$ as much money on January 1st as he does on January 2d; he has in the whole $\frac{4}{5}$ the amount required to pay his debts, and after paying out all his money, he still owes 28 dollars; how much does he receive January 1st, and how much January 2d?

23. Paid out $\frac{2}{5}$ of the money I had, and then borrowed $\frac{1}{2}$ as much as I had left; I had then 6

cents less than I had in the beginning; how much had I at first?

24. $\frac{5}{7}$ of 42 is $\frac{3}{8}$ of how many times $\frac{2}{3}$ of $13\frac{1}{2}$?

25. $\frac{3}{4}$ of 16 is $\frac{6}{7}$ of how many times $\frac{2}{5}$ of 15?

26. $\frac{4}{5}$ of 20 is $\frac{3}{8}$ of how many times $\frac{1}{7}$ of 56?

27. $\frac{5}{6}$ of 60 is $\frac{2}{5}$ of how many times $\frac{5}{8}$ of 40?

28. If a family of 6 persons will consume $2\frac{1}{4}$ barrels of flour in 6 months, what must be the number of persons in a family that will consume the same in $4\frac{1}{2}$ months?

29. $\frac{3}{4}$ of Henry's money is $\frac{5}{6}$ as much as Richard has, and twice Richard's money is $\frac{10}{9}$ as much as Charles has; if Charles has $42, how much each have Henry and Richard?

30. I have an orchard, in which $\frac{1}{6}$ of the trees bear peaches, $\frac{2}{3}$ bear apples, $\frac{1}{12}$ of them bear pears, 4 bear cherries, and 10 bear plums; how many trees in the orchard, and how many of each kind?

31. James spent $\frac{1}{3}$ of his money for a coat, $\frac{1}{12}$ for a vest, and $\frac{1}{4}$ for a railroad ticket; he loses $18, borrows $\frac{1}{8}$ as much as he had in the beginning, and has $8 left; how much had he at the beginning?

32. A man, on his way to market, was met by another man, who said: "Good-morrow, sir, with your hundred geese." Said he, "I have not a hundred, but if I had $\frac{1}{2}$ as many more and $2\frac{1}{2}$ geese, I should have a hundred." How many had he?

33. The head of a certain fish is 6 inches long, the tail is as long as the head and $\frac{1}{2}$ of the body, and the body is $3\frac{1}{2}$ times as long as the head; what is the length of the fish?

34. A man has 2 horses and a saddle; the saddle, which is worth $24, when put upon one horse is worth 3 times as much as the horse, but when put upon the other horse, is worth $\frac{1}{4}$ as much as the horse; what is the worth of each horse? What is the worth of each horse, with the saddle?

LESSON XL.

1. How much is 1 per cent. of $100? Of $10? Of $1?

NOTE. — Per cent. is a contraction of *per centum*, a Latin phrase, which means *by the hundred*. 1 per cent. means 1 of every hundred, or $\frac{1}{100}$; and 1 per cent. of $100 is $1, 1 per cent. of $10 or 100 dimes, is 10 cents or 1 dime, and 1 per cent. of $1 or 100 cents, is 1 cent.

2. How much is 1 per cent. of $2? Of $4? Of $7? Of $13? Of $64?

3. How much is 1 per cent. of $200? Of $300? Of $500? Of $800?

4. How much is 1 per cent. of $150? Of $250? Of $450? Of $750?

5. How much is 1 per cent. of $15? Of $150? Of $175? Of $480?

6. How much is 2 per cent. of $100? Of $10? Of $1?

7. How much is 2 per cent. of $30? Of $50? Of $48? Of $64?

8. How much is 3 per cent. of $320? Of $460? Of $280?

9. How much is 5 per cent. of 320 bushels?

10. How much is 4 per cent. of 225 tons?

11. How much is 6 per cent. of 100 pounds? Of 150 pounds? Of 362 pounds?

12. How much is 7 per cent. of 344 yards?

13. How much is 20 per cent. of $282? Of $560?

14. How much is 15 per cent. of $1000? Of $2500?

15. How much is 9 per cent. of 320 barrels?

16. A man divides his property among his four sons; to the youngest he gives 15 per cent., to the next 20 per cent., to the second 30 per cent., and

the remainder to the eldest; how much does he give the eldest?

NOTE. — The whole of any thing is of course 100 per cent. of itself.

17. If I give away 25 per cent. of my money to A, and 20 per cent..to B, and lend 55 per cent. to my brother, how much do I have left?

18. Bought a drove of cattle; in the first town I passed through, I sold 25 per cent. of them, and afterwards lost 8 per cent. of them; what per cent. of the drove had I then left?

19. If I draw 23 per cent. of my money out of the bank where it is placed, how much remains there?

20. If I borrow $70, and pay 6 per cent. a year for the use of it, how much do I pay in 6 years?

21. How much is 4 times 7 per cent. of $50?

22. How much is 8 times 6 per cent. of $80?

23. How much is 9 times 5 per cent. of $60?

24. How much is $\frac{1}{2}$ per cent. of $100? Of $10? Of $1?

25. How much is $\frac{3}{4}$ per cent. of $200? Of $20? Of $1?

26. How much is $5\frac{1}{3}$ times 3 per cent. of 150 pounds?

27. How much is 6 times 5 per cent. of 225 yards?

28. What is the commission for selling goods to the amount of $500, at $2\frac{1}{2}$ per cent. commission?

29. A city broker exchanged $400 on a country bank, at $\frac{1}{4}$ per cent.; what did he get for his trouble?

30. A man having $200 in uncurrent bank-bills, paid $1\frac{1}{2}$ per cent. to have them exchanged for current money; how much did he pay?

31. How much is $4\frac{1}{2}$ times 6 per cent. of 200? 10 times $5\frac{1}{2}$ per cent. of 100?

LESSON XLI.

1. What part of any thing is 10 per cent. ?

NOTE. — Since 1 per cent. is $\frac{1}{100}$, 10 per cent. is 10 times $\frac{1}{100} = \frac{10}{100} = \frac{1}{10}$.

2. What part of any thing is $12\frac{1}{2}$ per cent. ?
3. What part of any thing is 20 per cent. ?
4. What part of any thing is 25 per cent. ?
5. What fraction is 4 times 6 per cent. ?
6. What fraction is $5\frac{1}{2}$ times 8 per cent. ?
7. W   part of a ton is 24 per cent. of 75 per cent. ?   hat

NOTE. $\frac{24}{100}$ of 75 per cent. $= \frac{6}{25}$ of 75 per cent. $= 18$ per cent.

8. What part of a drove of sheep is 20 per cent. of 80 per cent. of it ?
9. What part of a ship is 25 per cent. of 50 per cent. of it ?
10. What per cent. of any thing is $\frac{1}{4}$ of it ?

NOTE. — Since 100 per cent. $=$ the whole of any thing, $\frac{1}{4}$ of any thing $= \frac{1}{4}$ of 100 per cent. $= 25$ per cent.

11. How many per cent. of any thing is $\frac{1}{5}$ of it ?
12. How many per cent. of any thing is $\frac{3}{5}$ of it ?
13. A man lost $\frac{1}{2}$ of his money ; what per cent. did he lose ?
14. If a man pays $\frac{3}{8}$ of his yearly income for board, what per cent. does he have left for other purposes ?
15. How many per cent. of any thing is $\frac{5}{8}$ of it ? $\frac{6}{25}$ of it ? $\frac{1}{4}$ of $\frac{3}{5}$ of it ?
16. In a certain school, 20 per cent. are in the first class, $\frac{2}{5}$ of the remainder in the second class, nd the rest in the third class, which has two equal

divisions; what fraction of the school is in each division?

17. What per cent. of 24 is 6?

NOTE. 6 is $\frac{6}{24}$ of $24 = \frac{1}{4}$ of 24; and $\frac{1}{4}$ of 100 per cent. $=$ 25 per cent.

18. What per cent. of 10 is 2?   Of 50 is 5?
19. What per cent. of 25 is 5?   10?   8?   20?
20. What per cent. of 42 is 7?   14?   21?   36?
21. 7 is what per cent. of 14?   Of 28?   Of 49? Of 70?
22. $25 is twice what per cent. of $150?
23. $\frac{1}{4}$ is what per cent. of $12\frac{1}{2}$?

NOTE. $12\frac{1}{2} = \frac{25}{2} = \frac{50}{4}$; and $\frac{1}{4}$ is the same per cent. of $\frac{50}{4}$ as 1 is of 50.

24. $\frac{1}{5}$ is what per cent. of 4?   Of 12?   Of 20?
25. 1 is what per cent. of $2\frac{1}{2}$?   Of $5\frac{2}{3}$?   Of $6\frac{1}{4}$?
26. $\frac{2}{3}$ of $\frac{3}{5}$ is what per cent. of $\frac{1}{6}$ of 12?
27. $\frac{1}{4}$ of 16 per cent. is what per cent. of 24 per cent.?
28. $\frac{1}{4}$ of 15 per cent. is what per cent. of 18 per cent.?
29. $\frac{3}{4}$ of 16 per cent. is what per cent. of 24 per cent.?
30. If a miller takes out 4 quarts for every bushel he grinds, what per cent. toll does he take?
31. If of a hogshead of sugar $\frac{1}{4}$ is sold, and of the remainder $\frac{1}{3}$ is rendered unsalable, what per cent. is the remainder?
32. $\frac{4}{5}$ of $\frac{1}{2}$ is what per cent. of $\frac{3}{4}$?
33. Sold, from a box of sugar containing 150 pounds, at one time 20 pounds, and at another time 30 pounds; what per cent. of the whole was sold?
34. Of 120 yards of cloth there have been sold 108 yards; what per cent. of the whole remains unsold?

LESSON XLII.

1. 12 is 6 per cent. of what number?

Note. — Since $\frac{6}{100} = 12$, $\frac{1}{100} = 2$, and $\frac{100}{100} = 100$ times 2 = 200.

2. 15 is 10 per cent. of what number?
3. 20 is 5 per cent. of what number?
4. $1\frac{1}{2}$ is 12 per cent. of what number?
5. $\frac{1}{4}$ is 7 per cent. of what number?
6. 24 is 40 per cent. of what number?
7. Sold rye so as to gain 25 cents a bushel, which was 20 per cent. of what it cost; what did it cost?
8. Paid 80 cents for making a vest, which was 20 per cent. of the cost of the cloth; what was the cost of the cloth?
9. Bought a horse for $160, which was 20 per cent. less than his true value; what was his true value?
10. If I buy a carriage for $228, at 24 per cent. less than what I can sell it for, what can I sell it for?

Note. 100 per cent. — 24 per cent. = 76 per cent., and since $\frac{76}{100} = 228$, $\frac{100}{100} = 300$.

11. A sells a lot of dry goods for $50, at 30 per cent. above cost; what did the goods cost him?
12. A sells B a watch, and gains 10 per cent.; B sells it again to C, and gains 20 per cent.; C pays $110; how much did it cost A?
13. Bought a cargo of flour, at 20 per cent. less than $7 a barrel, and sold it at 4 per cent. more than $7 a barrel; what per cent. was gained?
14. Sold 2 horses, at $200 apiece; on one there was a gain of 20 per cent., and on the other a loss

of 20 per cent.; was there a gain or a loss on the sale of the two, and of how much?

15. An apple-woman bought apples at 60 cents a hundred, and sold them at 1 cent apiece; how much per cent. did she gain?

16. When cloth is bought at $1.20 per yard, and sold at $1 per yard, how much per cent. is the loss?

17. Bought cloth at 80 cents per yard, and sold it at $1 per yard; how much per cent. is gained?

18. If a horse be bought for $80, and a cow for $25, and sold so as to gain 16 per cent., how much is received for them both?

19. Bought a horse for $200, which was 20 per cent. less than his worth, and sold him at 95 per cent. of his value; how much was gained?

20. If a barrel of flour is bought for $\frac{5}{8}$ of its market price, and sold for 4 per cent. more than the market price, how much per cent. is gained?

21. What per cent. does a merchant lose, who sells corn at 80 cents a bushel which cost him 90 cents?

22. Bought 12 barrels of flour for $110, and sold the same again for $9.50 a barrel; how much was the gain or loss?

23. I have a house which brings in $\frac{1}{4}$ of its value every 6 years; how much per cent. do I gain on it a year?

24. A bought goods for $100, and sold them to B so as to gain 10 per cent.; B sells them again to C, and gains 20 per cent.; how much did C pay? How much in money could A have made, if he had sold them himself to C, at the same price C paid?

25. A bought goods, and sold again to B at 10 per cent. more than he gave; B sells to C, and makes 20 per cent.; how many per cent. would A

make, were he to sell to C himself at the same price B got?

26. John Robinson buys a cargo of coal for $520, and sells it so as to gain 12 per cent.; for how much does he sell it?

27. Having a house worth $1400, I charge for it $9½ per month; how many per cent. a year do I get for it?

28. A grocer bought 100 eggs, at 15 cents a dozen, but 16 of them proved bad, and he sold the rest at 18 cents a dozen; how much per cent. did he gain?

29. Charles Thompson sells to John Johnson goods which he bought for $100, and gains 10 per cent.; he also sells some of the same goods to Peter Williams, and gains 30 per cent.; how much would Johnson gain if he were to sell all he bought to Williams at the same price Thompson sells to Williams?

30. If I sell to A, a retail dealer, and gain 10 per cent., and to B, who is one of A's customers, and gain 30 per cent., how much would A gain by selling to B at the same price as I sell to B?

31. If James has 50 per cent. more money than John, how many per cent. has John less than James?

32. Having a farm of 154 acres, worth $20 an acre, I let half of it at 5 per cent. a year on its value, and cultivated the other half myself, getting back $115 beyond all expenses; is it better for me to let my farm, or use it myself, and how much difference is there?

33. Two men go into partnership; A puts in $200, and B $150; their profits are $32; how much per cent. do they gain on their capital, and what is each one's share?

34. Buying $150 worth of goods, I lose 14 per

cent., but at the same time I buy $360 worth, and gain 9 per cent.; do I gain or lose on my whole capital, and how much?

35. A sells to B for $6 goods which he bought for $5; B sells again to C, and loses 10 per cent.; C sells to D at $7 what cost him $6; how much per cent. would A gain by selling directly to D?

36. A man gains 20 per cent., in each of three years, upon what he had at the beginning of the year; how much more then has he than when he began?

LESSON XLIII.

1. What is the interest of $1 or 100 cents for 1 year, at 6 per cent. by the year?

NOTE. — INTEREST is money paid for the use of money. It is reckoned at so many per cent., usually by the year. 6 per cent. means that $6 must be paid on every hundred dollars, 6 cents on every $1 or 100 cents, and so on. Hence, the interest on $1 or 100 cents for 1 year, at 6 per cent., = 6 cents.

2. If the interest of $1 for 1 year is 6 cents, what is the interest of $4 for 1 year? Of $5?

3. What is the interest of $1 for 2 years, at 6 per cent.?

NOTE. — Since the interest of $1 for 1 year, at 6 per cent., is 6 cents, for 2 years it will be 2 times 6 cents, which are 12 cents.

4. What is the interest of $1 for 3 years, at 6 per cent.? For 4 years? For 10 years?

5. What is the interest of $4 for 2 years, at 6 per cent.? Of $6? Of $8?

6. What is the interest of $10 for 3 years, at 6 per cent.? Of $12? Of $20? •

7. What is the interest of $50 for 1 year, at 6 per cent.? For 4 years?

8. What is the interest of $60 for 7 years, at 6 per cent.? For 9 years?

9. What is the interest of $1 for 1 year, at 6 per cent.? At 7 per cent.? At 5 per cent.?

10. What is the interest of $10 for 1 year, at 6 per cent.? At 7 per cent.? At 5 per cent.?

11. What is the interest of $100 for 1 year, at 6 per cent.? At 7 per cent.? At 5 per cent.?

12. What is the interest of $100 for 5 years, at 7 per cent.? For 6 years? For 7 years?

13. What is the interest of $150 for 2 years, at 7 per cent. For 3 years?

14. What is the interest of $106 for 3 years, at 7 per cent.? At 8 per cent.?

15. What is the interest of $320 for 4 years, at 10 per cent.? At 12 per cent.?

16. What is the interest of $415 for 5 years, at 7 per cent.? Of $540?

17. What is the interest of $100 for 1 month, at 6 per cent.?

Note.— Since the interest of $100 for 1 year, at 6 per cent., = $6, the interest for 1 month, or $\frac{1}{12}$ of a year, = $\frac{1}{12}$ of $6 = $ $\frac{6}{12} = $$\frac{1}{2} = 50$ cents.

18. What is the interest of $10 for 1 month, at 6 per cent.? Of $1? Of $6?

19. What is the interest of $60 for 2 months, at 6 per cent.? Of $9? Of $90?

20. What is the interest of $200 for 1 month, at 6 per cent.? Of $300? Of $370? Of $420? Of $426?

21. What is the interest of $100 for 4 months, at 6 per cent.? For 5 months? For 6 months? For 8 months? For 10 months?

22. What is the interest of $100 for 6 months, at 7 per cent.? At 8 per cent.?

23. What is the interest of $100 for 4 years

and 4 months, at 6 per cent. ? Of $200 ? Of $300 ? Of $50 ? Of $350 ?

24. What is the interest of $50 for 4 years and 6 months, at 7 per cent. ?

25. What is the interest of $150 for 6 years and 7 months, at 6 per cent. ?

26. What is the interest of $460 for 3 years and 2 months, at 5 per cent. ?

27. What is the interest of $625 for 5 years and 3 months, at 7 per cent. ?

LESSON XLIV.

1. What is the interest of $100 for 1 day, at 6 per cent. ? For 6 days ? For 15 days ?

NOTE. — In reckoning interest, it is customary to call 30 days a month. Hence, 1 day = $\frac{1}{30}$ of a month ; 6 days = $\frac{6}{30}$ = $\frac{1}{5}$ of a month ; $\frac{15}{30}$ = $\frac{1}{2}$ of a month, and so on. Since the interest of $100 for 1 month is 50 cents, for $\frac{1}{30}$ of a month it is $\frac{1}{30}$ of 50 cents = $1\frac{2}{3}$ cents ; for $\frac{1}{5}$ of a month it is $\frac{1}{5}$ of 50 cents = 10 cents, and for $\frac{1}{2}$ of a month it is $\frac{1}{2}$ of 50 cents = 25 cents.

2. What is the interest of $1 for 30 days, at 6 per cent. ? For 60 days ? For 120 days ? For 20 days ? For 40 days ?

3. What is the interest of $20 for 5 days, at 6 per cent. ? For 12 days ? For 15 days ?

4. What is the interest of $50 for 21 days, at 7 per cent. ? For 24 days ? For 45 days ?

5. What is the interest of $40 for 4 years, at 6 per cent. ? For 4 months ? For 4 years and 4 months ? For 20 days ? For 4 years, 4 months and 20 days ?

6. What is the interest of $400 for 6 years, 2 months and 10 days, at 2 per cent. ?

7. What is the interest of $100 for 4 months and 15 days, at 7 per cent. ?

8. What is the interest of $125 for 4 years, 2 months and 10 days, at 7 per cent.?

9. What is the interest of $120 for 3 years, 4 months and 10 days, at 6 per cent.?

10. What is the interest of $140 for 1 year and 20 days, at 6 per cent.?

11. What is the interest of $30 for 6 years, at 5 per cent.? At 7 per cent.?

12. What is the interest of $10 for 4 years and 8 months, at 7 per cent.? At 8 per cent.?

13. What is the interest of $150 for 5 years and 4 months, at 6 per cent.?

14. What is the interest of $320 for 2 years, 4 months and 20 days, at 6 per cent.?

15. What is the interest of $45 for 4 years, 8 months and 16 days, at 6 per cent.?

16. What is the interest of $80 for 6 years, 6 months and 12 days, at 6 per cent.?

17. Required the interest of $126 for 3 years, 6 months and 20 days, at 6 per cent.

18. Required the interest of $175 for 4 years, 4 months and 18 days, at 6 per cent.

19. Find the interest of $210 for 3 years and 4 months, at 7 per cent.

NOTE.—In reckoning interest at 4, 5, 7, or 8 per cent., it is often most convenient to find the interest at 6 per cent., and then change it to the given per cent. The interest at 4 per cent. is equal to that at 6 per cent. less $\frac{1}{3}$ of itself; 5 per cent. is 6 per cent. less $\frac{1}{6}$ of itself; 7 per cent. is 6 per cent. with $\frac{1}{6}$ added; and 8 per cent. is 6 per cent. with $\frac{1}{3}$ added.

20. Find the interest of $42 for 3 years and 2 months, at 6 per cent. At 4 per cent. At 5 per cent. At 7 per cent. At 8 per cent.

21. Find the interest of $85 for 2 years, 8 months and 12 days, at 6 per cent. At 4 per cent. At 5 per cent. At 7 per cent. At 8 per cent.

22. What is the interest of $160 for 3 years, 4 months and 9 days, at 6 per cent.? At 5 per cent.? At 8 per cent.? At 7 per cent.?

23. What is the interest of $200 for 2 years, 6 months and 15 days, at 4 per cent.? At 6 per cent.?

24. The interest of $240 for a certain time is $18, at 6 per cent.; how much is it at 8 per cent.?

25. The interest of $225 for 4 years is $54; what is the interest for 1 year? For 3 months? For 2 months? For 1 month?

26. The interest of $120 for 1 year is $7.20; how much is it for 4 months? For 2 months? For 1 month and 10 days? For 20 days?

LESSON XLV.

1. What is the amount of $40 for 2 years, at 6 per cent.?

NOTE. — The amount is the sum at interest, or the principal, and the interest taken together. The interest on $40 for 2 years, at 6 per cent. = $4.80; and $40+$4.80 = $44.80, the amount.

2. What is the amount of $60 for 4 years, at 4 per cent.?

3. What is the amount of $80 for 3 years, at 5 per cent.?

4. What is the amount of $100 for 6 years, at 4 per cent.?

5. What is the interest of $70 for 6 years, at 7 per cent.?

6. What is the amount of $125 for 2 years, at 7 per cent.?

7. What is the interest of $100 for 3 years and 4 months, at 7 per cent.?

8. What is the interest of $25 for 6 years and 1 month, at 6 per cent.?

9. What is the interest of $10 for 4 years and 2 months, at 7 per cent. ?

10. What is the amount of $150 for 4 years and 1 month, at 7 per cent. ?

11. Required the amount of $200 for 2 years and 2 months, at 4 per cent.

12. Required the amount of $145 for 6 years and 3 months, at 8 per cent.

13. Required the amount of $400, at $3\frac{1}{2}$ per cent., for 2 years, 2 months and 20 days.

14. What is the amount of $140 for 5 years, 4 months and 10 days, at 7 per cent. ?

15. What is the amount of $200 for 2 years, 6 months and 14 days, at 6 per cent. ?

16. What is the interest of $15 for 6 years, 4 months and 2 days, at 8 per cent. ?

17. What is the amount of $360 for 8 years, 2 months and 12 days, at 6 per cent. ? .

18. What is the amount of $180 for 6 years, 7 months and 13 days, at 7 per cent. ?

19. What is the amount of $200 for 5 years and 9 months, at 4 per cent. ?

20. What is the amount of $140 for 3 years, 8 months and 18 days, at 8 per cent. ?

21. What is the amount of $175 for 4 years, 4 months and 9 days, at $6\frac{1}{2}$ per cent. ?

22. What is the amount of $140 for 3 years, 6 months and 18 days, at 3 per cent. ?

23. Required the amount of $160 for 9 years, at 5 per cent.

24. Required the amount of $300 for 5 years, 8 .months and 6 days, at 6 per cent.

25. What is the amount of $250 for 3 years, 6 months and 24 days, at 7 per cent. ?   .

26. What is the amount of $500 for 1 year, 5 months and 20 days, at 10 per cent. ?

LESSON XLVI.

1. If the interest of $120 for 1 year is $9, what is the interest of $60? Of $20? Of $10?

2. If the interest of $400 for 1 year is $24, what is the interest of $100? Of $10? Of $1?

3. If the interest of $12 for 5 years is $3.60, what is the interest of $6? Of $60? Of $300?

4. If the interest of $160 for 2 years is $18, what is the interest of $80? Of $40?

5. If the interest of $280 for 3 years is $46, what is the interest of $70? Of $140?

6. If the interest of $100 for 3 years is $20, what will it be for 6 years? For 10 years? For 4 years? For 1 year?

7. If the interest of $180 for 2 years is $23, what is it for 3 years? For 7 years? For 1 year?

8. If the interest of $200 for 3 years is $27, what is its interest for 1 year? For 1 month? For 4 months? For 7 months?

9. If the interest of $200 for 1 month is $1, what is it for 15 days? For 5 days? For 1 day?

10. If the interest of $140 for 1 year and 4 months is $10, what is it for 8 months? For 4 months? For 2 months? For 1 month?

11. The interest of $240, at 6 per cent., is $27; what is it at 3 per cent.? At 1 per cent.?

12. The interest of $320, at 7 per cent., is $15; what is it at 1 per cent.? At 8 per cent.?

13. If the interest of $100 for 3 years is $21, what is the rate per cent.?

14. If the interest of $300 for 1 year is $21, what is the rate per cent.?

15. What is the interest of $100 for 4 years, at 1 per cent.? If the interest of $100 for 4 years is $20, what is the rate per cent.?

16. If the interest of $300 for 2 years is $48, what is the rate per cent.?

NOTE. — Since the interest of $300 for 2 years, at 1 per cent., is $6, $48 will be as many per cent. as $48 contains times $6.

17. If the interest of $200 for 2 years is $12, what is the rate per cent.?

18. If the interest of $400 for 3 years is $24, what is the rate per cent.?

19. If the interest of $150 for 3 years and 4 months is $35, what is the rate per cent.?

20. If the interest of $25 for 1 year and 8 months is $1, what is the rate per cent.?

NOTE. — The interest of $25 for 1 year and 8 months, at 1 per cent., is $1\frac{2}{3}$ times $\frac{1}{100}$ of $25 = \frac{1}{60}$ of $25 = \$ \frac{5}{12}$, and the rate per cent. required will be as many per cent. as $\$ \frac{5}{12}$ is contained times in $1, which are $\frac{12}{5} = 2\frac{2}{5}$ times. Therefore, $2\frac{2}{5}$ per cent. is the answer.

21. If the interest of $50 for 1 year and 3 months is $3, what is the rate per cent.?

22. If the interest of $75 for 4 years and 2 months is $12, what is the rate per cent.?

23. A man paid $8 for the use of $48 for 1 year and 4 months; what was the rate per cent.?

24. John Niles lends Harry Hubbard $30 for 2 years and 6 months; Harry has to pay at the end of the time $36; what is the rate per cent.?

NOTE. — The principal, $30, has to be paid back, and is of course a part of the $36; the remainder, $6, is the interest, by which the rate per cent. is to be found.

25. A note of $100, being on interest 2 years and 2 months, amounted to $125; what was the rate per cent.?

26. A gentleman lent $60 for 1 year and 6 months, and received $90; what was the rate per cent.?

LESSON XLVII.

1. What principal, at 10 per cent., is sufficient to gain $6 in 4 years?

NOTE. — Since the interest of $1 for 4 years, at 10 per cent., is 40 cents, $6 is the interest of as many dollars as 40 cents or 4 dimes is contained times in $6 or 60 dimes, which are 15; therefore, $15 is the answer.

2. What principal, at 4 per cent., is sufficient in 2 years to gain $8?

3. What is the principal that, in 4 years, at 3 per cent., will gain $6?

4. What principal, at 6 per cent., is sufficient in 5 years to gain $10?

5. What principal is sufficient in 2 years and 4 months, at 4 per cent., to gain $64?

6. What principal is sufficient in 2 years, at 8 per cent., to gain $10?

7. What principal is sufficient in 6 years, at 1 per cent., to gain $20?

8. What principal, at 4 per cent., in 1 year and 7 months, will gain $5?

9. What principal, in 6 years and 4 months, at 6 per cent., will gain $18?

10. What principal is sufficient in 7 years to gain $18, at 7 per cent.?

11. The interest on a note for 4 years and 2 months, at 4 per cent., was $60; what was the principal?

12. If the interest of $50, at 6 per cent., is $6, how long was it on interest?

NOTE. — Since $50 at 6 per cent. will require 1 year to gain $3; it will require as many years to gain $6 as $3 is contained times in $6, which are 2 times. Therefore, 2 years is the answer.

13. If the interest of $60, at 2 per cent., is $12, how long has it been on interest?

14. How long must $100 be on interest, at 4 per cent., to gain $40?

15. A note of $80, being on interest at 8 per cent., amounted to $160; how long was it on interest?

16. How long must $10 be on interest, at 5 per cent., to gain $3?

17. If the interest of $20, at 4 per cent., is $4, how long a time has it been on interest?

18. If the interest of $50, at 8 per cent., is $12, how long has it been on interest?

19. Required the time that $40 must be on interest, at 2 per cent., to gain $8.

20. A gentleman lent $60, at 6. per cent., and received $140; how long was it on interest?

21. A sum of money is on interest at 6 per cent., how long will it take it to double itself?

NOTE. — That is, if it gain 6 per cent. a year, how long will it take to gain 100 per cent.?

22. How long will it take a sum of money to double itself, at 9 per cent.? At 18 per cent.?

23. A given principal gains $\frac{1}{4}$ of $\frac{3}{7}$ of itself a year; how long will it take it to double itself?

NOTE. $\frac{1}{4}$ of $\frac{3}{7} = \frac{3}{28}$; if the principal gains $\frac{3}{28}$ of itself in one year, it will take it as many years to gain as much as itself as $\frac{3}{28}$ is contained times in $\frac{28}{28}$, which gives $\frac{28}{3} = 9\frac{1}{3}$ years.

24. A given principal gains $\frac{2}{3}$ of $\frac{1}{5}$ of itself a year; how long will it take to double itself? To gain $\frac{1}{2}$ of itself? $\frac{1}{4}$ of itself?

25. If I loan $1200, at 8 per cent. a year, how long will it be in gaining $80?

26. If I loan $500, at 7 per cent. a year, how long will it be in gaining $350?

LESSON XLVIII.

1. What principal, in 4 years, at 5 per cent., will amount to $96?

NOTE. — The interest of any sum of money at 5 per cent., for 4 years, is $\frac{20}{100}$ or $\frac{1}{5}$ of the principal; then the amount is $\frac{6}{5}$ of the principal, for the amount is $\frac{5}{5}$ of the principal, plus the interest, $\frac{1}{5}$; then, since $\frac{6}{5}$ of the principal = $96, the principal = $\frac{5}{6}$ of $96 = $80.

2. What is the present worth of $96, due 4 years hence, at 5 per cent.?

NOTE. — The present worth of a sum due at some future time is what should be paid now, instead of paying that sum then; it is, therefore, equivalent to a principal which, being put at interest, will amount to the debt at the time of its becoming due.

3. What is the present worth of $50, due in 5 years, at 5 per cent.?

NOTE. — That is, what principal, put at interest for 5 years, at 5 per cent., will amount to $50?

4. What is the present worth of $136, due in 6 years, at 6 per cent.?

5. What is the present worth of $172, due in 9 years, at 8 per cent.?

6. What is the present worth of $7½, due in 5 years, at 10 per cent.?

7. What is the present worth of $78, due in 1 year, at 4 per cent.?

8. What is the present worth of $85, due in 10 years, at 7 per cent.?

9. What is the present worth of $96, due in 8 years, at 7½ per cent.?

10. What is the discount on $100, due in 5 years, at 5 per cent.?

NOTE. — Discount is the allowance or deduction made for paying money before it is due; it is equivalent to the interest on the present worth of the debt up to the time of the same becoming payable. The discount, therefore, may be found by subtracting the present worth from the given sum or debt.

11. What is the discount on $84, due in 8 years, at 5 per cent.?

12. What is the discount on $77, due in 9 years, at 6 per cent.?

13. What is the discount on $66, due in $6\frac{1}{2}$ years, at 10 per cent.?

14. What is the discount on $74, due in 6 years, at 8 per cent.?

15. What is the discount on $100, due in 2 years, at 7 per cent.?

16. What is the discount on $77⅝, due in 4 years, at 10 per cent.?

17. What is the discount on $57, due in 2 years, at 7 per cent.?

18. What is the discount on $63, due in 10 months, at 6 per cent.?

19. What is the discount on $75, due in 1 year and 4 months, at 6 per cent.?

LESSON XLIX.

1. Divide a sum of money between A and B, giving A $3 as often as you give B $2; what share of the money will each receive?

NOTE. $3 + $2 = $5; and $3 = $\frac{3}{5}$ of $5, and $2 = $\frac{2}{5}$ of $5. Therefore, A receives $\frac{3}{5}$, and B $\frac{2}{5}$ of the money.

2. If you divide $15 between A and B, by giving A $3 as often as you give B $2, how many dollars will each receive?

3. Divide $35 between John Wilson and Simon Edwards; how much will each get, if John receive $4 as often as Simon receives $3?

4. A man has two sons, to whom he leaves $660; to the elder $7 to every $4 he leaves the younger; how much does each get?

5. A gentleman has 3 sons, to whom he gives 98 cents to celebrate the Fourth of July; Edward, the eldest, has 2 times as much as Robert, the second, and 4 times as much as John, the youngest: how much does each get?

6. Two men buy a cask of beer, containing 30 gallons; one pays $4, and the other $3½; what part does each get?

7. Divide 55 into two parts, of which the larger is $\frac{6}{5}$ of the less; how great is each?

Note.—The larger is 6 times $\frac{1}{5}$ the less, and the smaller 5 times $\frac{1}{5}$ of itself; therefore, 55 must equal 6 + 5, or 11 times $\frac{1}{5}$ the less. $\frac{1}{5}$ of the less is then 5, the greater is 30, and the less 25.

8. Divide 27 into two parts, of which one is $\frac{4}{5}$ of the other.

9. I go into partnership with Thomas Gould, and pay in $5 of capital to every $4 he pays; how much more do I pay than he? How much less does he than I? What share of the whole do I pay, and what share does he pay? If we have a profit of $63, how much ought I to get? How much ought Gould?

10. C and D hire a pasture together; C pays $12, and D $10; C puts in 6 cows; how many should D put in?

11. John Stevens and Samuel Judkins hire a pasture together; John puts in 3 cows for 8 days, and Samuel 4 cows for 7 days; they pay $5 rent; what part of it should each pay?

12. While travelling, I met 11 beggars; 2 were cripples, 3 were blind, and 6 were too lazy to work; I gave each cripple 2 times as much as each blind

man, and each blind man 3 times as much as each lazy man; in all I gave away 54 cents; how much did each cripple receive?

13. Mary had 40 apples; she gave $\frac{2}{5}$ to her schoolmates, and divided the rest between her two sisters and herself, taking only $\frac{1}{5}$ as many as both her sisters; how many did she have for herself?

14. In a granary there is twice as much rye as wheat, twice as much wheat as buckwheat, and $\frac{1}{3}$ as much barley as rye; there are 120 bushels in all; how much of each kind?

15. John has 26 cents' worth of marbles, $\frac{3}{5}$ of the number of which are worth 8 to a cent, $\frac{3}{10}$ 2 to a cent, and the remainder 1 cent apiece; how many has he?

16. I went to the city with $5\frac{1}{2}$ in my purse; I spent $\frac{1}{10}$ of it in paying my fare, $\frac{1}{8}$ of the remainder for a reading-book, and bought with what was left twice as many grammars at 20 cents apiece, as I did spelling-books at 15 cents apiece; how many did I buy of each?

17. If $\frac{5}{8}$ of a bushel of potatoes cost $\$\frac{1}{8}$, how much will $\frac{2}{5}$ of a bushel cost?

18. Two men entered into partnership; the first put in $500 for $4\frac{1}{5}$ months, and the second $600 for $3\frac{1}{8}$ months; their profits are $110; how much ought each to have?

19. I have one book which has 16 pages to a sheet, and another which has 36; they have each the same number of pages, and together 39 sheets; how many pages has each?

20. A, B and C, enter into partnership; A puts in $\frac{1}{4}$, B $\frac{2}{5}$, and C the remainder; after a while A withdraws his, and the capital now is $480; how much does each put in?

21. Divide 69 into two parts, which shall be to each other as $\frac{5}{8}$ to $\frac{4}{5}$.

22. Three men hired a pasture for $60; A put in 4 oxen, B 3 oxen, and C 5 oxen; how much ought each to pay?

23. Two men start from New York, and travel in opposite directions, one at the rate of $4\frac{1}{4}$ miles an hour, and the other at the rate of $6\frac{3}{4}$ miles an hour; how far apart will they be at the end of 1 hour? How far in 4 hours? 6 hours? 10 hours?

24. A gentleman wished to give $100 for benevolent objects; having given away $\frac{1}{2}$, $\frac{1}{4}$ and $\frac{1}{8}$ of it, he divided the rest equally among 5 poor widows; how much did each widow receive?

25. A man having returned from California, being asked how much money he had made, answered, that if he had made as much more, and half as much more, he should have $1000; how much had he made?

LESSON L.

1. A cistern is filled by a pipe in $4\frac{1}{2}$ hours; what part of it is filled in 1 hour?

2. A man can do $\frac{1}{5}$ of a piece of work in 1 hour; how long will it take him to do the whole?

3. Two men together can eat $\frac{1}{20}$ of a barrel of crackers in 1 day; how long will it take them to eat the whole?

4. A can do a piece of work in 8 days, and B the same work in 12 days; how long will it take them both?

Note. — Since A can do $\frac{1}{8}$ of the work in 1 day, and B $\frac{1}{12}$, both can do $\frac{1}{8} + \frac{1}{12}$, or $\frac{3}{24} + \frac{2}{24} = \frac{5}{24}$ in 1 day; therefore, it will take them both as many days to do the whole work as $\frac{5}{24}$ is contained times in $\frac{24}{24}$, which gives $\frac{24}{5} = 4\frac{4}{5}$ days.

5. Two men can each do $\frac{2}{7}$ of a piece of work

10

in 1 day; how long will it take them both to do the whole?

6. Two men set out to mow a field; the first can mow it in 16 days, and the second in 20; how long will it take them both?

7. A cistern has 3 pipes; the first will fill it in 2 hours, the second in 3 hours, and the third in 6 hours; how long will it take them all to fill it?

8. A cistern has 2 pipes; the first will fill it in 2 hours, and the second will empty it in 3 hours; if both pipes are open, how long will it take to fill the cistern?

9. A cistern has 4 pipes; the first will fill it in 2 hours, the second will fill it in 3 hours, the third in 4 hours, and the fourth will empty it in 2 hours; now, if the pipes are all open at the same time, how long will it take to fill the cistern?

10. A and B together can build a wall in 16 days, but with the aid of C they can build it in 10 days; how long will it take C to build it alone?

11. A can reap a certain piece of rye in $\frac{2}{5}$ of a day, B in $\frac{3}{4}$ of a day, and C in 1 day; how long will it take them together to finish the piece, after C has been reaping $\frac{1}{2}$ of a day?

12. A can cut a cord of wood in $\frac{2}{3}$ of a day, B in $\frac{3}{4}$ of a day, and C in $\frac{2}{3}$ of a day; after A and B have cut $\frac{1}{4}$ of a day, how long will it take C to finish the cutting of the remainder of the cord?

13. Divide 24 into 2 parts, which shall be to each other as $1\frac{2}{3}$ is to 3.

14. Divide 36 into 3 parts, which shall be respectively as $2\frac{1}{3}$, $1\frac{3}{4}$, and $1\frac{1}{4}$.

15. A and B can do a piece of work in 15 days, and B alone in 24 days; how long would it take A alone?

16. Three men go into partnership; the first pays $7 as often as the second pays $4, and the

third $5 as often as the second pays $8; they all pay in $900; how much does each pay?

17. Divide 86 into two parts, of which the larger is 5¼ times as much as the less.

18. Divide 50 into three parts, of which the first is ⅔ as large as the second, and the third 3 times as large as the first and second.

19. Three men, A, B and C, are to share $400 in the proportion of ½, ¼ and ⅛ respectively; but, as C died, it is required to divide the whole sum properly between the other two; how much should each receive?

20. If 4 pounds of flour will make 40 four-cent loaves of bread, how many six-cent loaves can be made from the same quantity?

21. A man, in distributing some money to several indigent persons, gave $2½ to one man, $3¼ to another, $4¾ to another, $5¼ to another, and $1⅜ to another; how many dollars did he give away?

22. A student, having a bible, a dictionary, and an algebra upon his table, was asked the price of each; he answered that his bible cost twice as much as his dictionary, the dictionary cost twice as much as the algebra, and that the three books cost $10; what was the cost of each book?

LESSON LI.

1. If $6 worth of provisions will last 5 men 7 days, how long will they last 10 men?

NOTE. — What will last 5 men 7 days, will last 1 man 5 times 7 days, or 35 days; and what will last 1 man 35 days, will last 10 men $\frac{1}{10}$ of 35 days, or 3½ days.

2. If a barrel of meat will last 12 men 20 days, how long will it last 16 men? 30 men? 40 men?

3. If a barrel of beer last 5 men 16 days, how long will it last 8 men? How many men will it last 8 days?

4. If 3 men can mow a field in '10 hours, how long will it take them, if 2 men be added to their number?

5. If 7 horses consume 16 tons of hay a year, how many tons do 5 horses consume?

6. If 12 men can dig a well in 4 days, in how many days can 15 men dig it?

7. If 8 bushels of grain will last 7 horses 5 days, how long will 16 bushels last 4 horses?

NOTE. — If 8 bushels last 7 horses 5 days, they will last 4 horses $\frac{7}{4}$ of 5 days, or $8\frac{3}{4}$ days ; and if 8 bushels last $8\frac{3}{4}$ days, 16 bushels will last 2 times as long, since 16 bushels are 2 times 8 bushels ; and 2 times $8\frac{3}{4}$ days are $17\frac{1}{2}$ days.

8. If a ton of hay will last 8 cows 7 days, how long will it last 11 cows?

9. If a ton of hay will last 7 cows 8 days, how many cows will it last 14 days?

10. If $300 gain $12 in 8 months, what sum would it require to gain $8 in 2 months?

11. If $100 gain $6 in 12 months, how many months would it require for $400 to gain $10?

12. If 40 bushels of oats are sufficient for 5 horses 6 weeks, how many bushels would it require to supply 15 horses 8 weeks?

13. If 20 oxen eat 4 tons of hay in 30 days, how many oxen would it take to eat 12 tons in 60 days? In 50 days? In 75 days?

14. If 7 men can build 8 rods of wall in 4 days, how many days would it take 14 men to build 32 rods of wall? 40 rods? 80 rods?

15. If 12 men can reap a field of 4 acres in 8 days, by laboring 6 hours a day, how many acres would 6 men reap in 12 days, by laboring 7 hours per day?

16. If the interest of $250 for 10 months is $ 12½, what is the interest of $450 for 5 months?

17. If 2 oxen, or 3 cows, will eat 3 tons of hay in 18 weeks, how much hay will 6 oxen and 1 cow eat in 9 weeks?

18. There are in a fort 200 men, with provisions sufficient to last 6 months; how many must leave after the provisions are half gone, that the remaining men shall have just sufficient for 6 months?

19. If 8 men can dig a ditch 30 rods long in 20 days, how long will it take 10 men to dig a ditch 15 rods long?

20. A workman, laboring 10 hours a day, will build in 20 days 30 rods of wall; how long will it take 10 workmen, laboring 9 hours a day, to build 45 rods?

21. If 6 men can do the work of 24 women, and 4 women do the work of 6 boys, how many men can do the work of 18 boys?

NOTE. — Since 6 men can do the work of 24 women, 1 man can do the work of 4 women, and as 4 women can do the work of 6 boys, 1 man can do the work of 6 boys, and 8 men the work of 18 boys.

22. If 5·pounds of cheese are equal in value to 2 pounds of butter, and 6 pounds of butter to 2 bushels of corn, how many pounds of cheese will pay for 4 bushels of corn?

23. If the relative value of oak wood to spruce is as 2 to 1, and that of spruce to pine as 7 to 8, how many cords, composed of spruce and pine in .equal parts, will equal 10 cords of oak?

24. If 3 men can build a boat in 12 days, when the days are 12 hours long, how long will it take 5 men to build the same, when the days are 10 hours long?

25. If $150 gain $9 in 12 months, in what time will $200 gain $18?

LESSON LII.

NOTE TO TEACHERS. — Lessons LII. - LXIII., or any portion, if deemed too difficult, may be omitted, without greatly impairing the unity of the work.

1.   Add 38 and 46 together.*

NOTE. 38 is 3 tens and 8; 46 is 4 tens and 6; 38 and 46 is then 3 tens and 4 tens, or 7 tens, and 8 and 6, or 14; 14 is 1 ten and 4; 38 and 46 is then 7 tens and 1 ten and 4, or 8 tens and 4, that is 84.

2.   What is the sum of 23 and 25?   29 and 33?
3.   What is the sum of 57 and 34?   76 and 18?
4.   How many are 37 and 39?   62 and 28?
5.   How many are 51 and 27?   26 and 67?
6.   How many are 68 and 54?

NOTE. 6 tens and 5 tens are 11 tens; 10 and 4 are 14, or 1 ten and 4; 11 tens and 1 ten and 4 are 12 tens and 4, or 124.

7.   How many are 66 and 67?   58 and 45?
8.   How many are 93 and 14?   56 and 65?
9.   How many are 83 and 36 and 68?
10.   How many are 95 and 21 and 70?
11.   How many are 54 and 13 and 28?
12.   How many are 84 and 74 and 21?
13.   How many are 13 and 63 and 45?
14.   How many are 108 and 22 and 36?
15.   How many are 53 and 63 and 75?
16.   How many are 28 and 31 and 64?
17.   How many are 91 and 26 and 67?
18.   How many are 34 and 45 and 77?
19.   How many are 105 and 28 and 43?
20.   How many are 185 and 36 and 22?
21.   How many are 47, 38, 75 and 22?
22.   How many are 54, 68, 36 and 25?
23.   How many are 112, 27, 39 and 78?

---

* ADDITION is the process of collecting two or more numbers into *one sum*. The result is called the *amount*.

24. How many are 36, 25, 34 and 126?
25. How many are 203, 37 and 95?
26. How many are 81, 181, 243 and 6?
27. How many are 68, 84, 27 and 126?
28. How many are 58, 95 and 137?
29. How many are 154, 46 and 100?
30. How many are 130, 67 and 255?
31. How many are 25, 314 and 42?
32. How many are 29, 137, 62 and 38?
33. How many are 164, 212 and 367?
34. How many are 540, 24 and 365?
35. How many are 330, 45 and 127?
36. How many are 300, 130 and 76?
37. How many are 550, 28 and 164?
38. How many are 2000, 1500, 370 and 28?
39. How many are 8000, 1200, 140 and 4?
40. How many are 6300, 450 and 76?
41. How many are 3900, 460 and 68?

LESSON LIII.

1. Add 7, 19, 20, 3, 4, 6, 8, 17 and 22.
2. Add 3, 11, 26, 4, 15, 27, 24, 12 and 8.
3. Add 1, 5, 7, 9, 21, 13, 6, 8 and 14.
4. Add 11, 8, 2, 3, 18, 5, 24 and 1.
5. Add 13, 45, 61, 28, 45, 18, 36 and 63.
6. Add 1, 4, 6, 11, 21, 31 and 8.
7. Add 6, 25, 50, 1, 38, 16, 11 and 41.
8. Add 7, 40, 9, 3, 8, 21 and 19.
9. Add 5, 57, 43, 21, 6, 8 and 14.
10. Add 6, 42, 11, 56, 25, 50, 1, 20, 50 and 13.
11. Add 7, 13, 16, 23, 25, 41, 38 and 27.
12. Add 16, 93, 12, 17, 24, 36 and 4.
13. Add 6, 81, 93, 21, 7, 36, 40 and 28.
14. Add 7, 99, 19, 29, 40, 27 and 28.
15. Add 8, 89, 40, 25, 16, 9, 4 and 1.
16. Add 8, 25, 70, 56, 8, 31, 22 and 64.

17. Add 7, 98, 9, 5, 4, 32, 20 and 17.
18. Add 75, 39, 56, 28, 54, 91 and 8.
19. Add 39, 37, 59, 20, 25, 41 and 53.
20. Add 58, 24, 63, 71, 55, 11, 7 and 25.
21. Add 36, 28, 63, 50, 55, 25, 92 and 86.
22. Add 11, 67, 75, 20, 91, 51, 46, 62 and 77.
23. Add 38, 34, 27, 31, 45, 52, 96 and 27.
24. Add 50, 36, 25, 12, 58, 71, 26, 25, 42, 50, 36 and 68.
25. Add 8, 26, 56, 25, 33, 50, 76, 22, 42, 96 and 98.
26. Add 6, 22, 42, 2, 25, 56, 23, 5, 29, 31, 30 and 41.
27. Add 136, 28, 29, 54, 151, 93, 27, 6, 8, 36 and 16.
28. Add 131, 328, 464 and 22.
29. Add 118, 211, 313, 414 and 515.
30. Add 117, 293, 360 and 404.
31. Add 312, 266, 313, 427 and 195.
32. Add 111, 34, 147, 272 and 36.
33. Add 316, 424, 686 and 332.

LESSON LIV.

1. Subtract 33 from 76.*

Note. 76 is 7 tens and 6 ; 88 is 3 tens and 3; 3 tens from 7 tens leaves 4 tens, and 8 from 6 leaves 8 ; the difference required is, therefore, 4 tens and 3, or 43.

2. Take 28 from 99 ; 33 from 85 ; 46 from 78.
3. Take 54 from 97 ; 5 from 98 ; 54 from 89.
4. Take 8 from 42.

---

* SUBTRACTION is the process of taking one number from another, to find the difference. The larger of the two given numbers is called the *minuend*, the less number the *subtrahend*, and the result, or the number found, the *difference*, or *remainder*.

NOTE. — Since 8 cannot be taken from 2, we may take a part of the 40 to help us ; but as 10 is more than 8, *one* ten and 2 is enough, and it is easier to take one *ten* than any other number from 4 tens ; besides, we have even tens left for the first figure . of the answer. Therefore, 8 from 42 is the same as 3 tens and 1 ten and 2, less 8 ; 1 ten and 2, or 12 less 8, is 4 ; then the answer is 3 tens and 4, or 34.

5. Take 6 from 53 ; 9 from 64 ; 7 from 69.
6. Take 5 from 30 ; 8 from 81 ; 2 from 77.
7. Take 1 from 20 ; 2 from 30 ; 3 from 40.
8. Take 4 from 50 ; 5 from 60 ; 7 from 70.
9. Take 6 from 80 ; 7 from 90 ; 9 from 100.
10. Take 8 from 85 ; 6 from 74 ; 9 from 95.
11. Take 28 from 76.

NOTE. — Since 8 cannot be taken from 6, 76 may be called 6 tens and 16 ; 8 from 16 leaves 8, and 2 tens from 6 tens leaves 4 tens ; therefore, 4 tens and 8, or 48, is the answer.

12. Take 37 from 100 ; 29 from 73 ; 56 from 195.
13. Take 39 from 273 ; 126 from 285 ; 353 from 576.
14. Take 229 from 310 ; 292 from 371 ; **254** from 376.
15. Take 95 from 511 ; 128 from 464 ; 367 from 633.
16. Add 232 and 328, and subtract 414.
17. How many are 365 and 147, less 228 and 195?
18. How many are 424 and 181, less 564 ?
19. Add 3480 and 225, and take 1009 from their sum.
20. Add to 252 less 178, 341 less 279.
21. Add 254 and 316, and subtract 142.
22. Add 573, 96, 69, 251, 224 and 197.
23. Add 385, 377, 451 and 686.
24. Subtract 285 from 269 and 143.
25. Add 676 and 343 to 216 less than 435.
26. Subtract 285 from 676 more than 1020.

11

27. Add 6000, 2400, 560 and 79, and subtract from their sum 2500 and 1636.

28. Add together 2500, 370 and 45, and subtract from their sum 282 and 1500.

29. Find the sum of 5200, 380 and 76, from which take the sum of 1800 and 677.

30. Thomas Brown buys of George Jones, at different times, goods to the amount of $38.50, $54.25, and $24.60; after this he pays $90, and buys more goods to the amount of $15.70; how much does he now owe?

LESSON LV.

1. Multiply 18 by 3.*

NOTE. 18 is 10 and 8; 8 times 10 are 8 tens, or 80, and 8 times 8 are 24, or 2 tens and 4; then 8 times 18 are 80 and 20 and 4, or 50 and 4, or 54.

2. Multiply 16 by 5; by 6; by 4.
3. How many are 8 times 11? 8 times 12?
4. How many are 8 times 13?

NOTE. 8 times 10 are 8 tens, and 8 times 3 are 2 tens and 4; 8 tens and 2 tens are 10 tens, or 100.

5. Multiply 18 by 7; 19 by 8; 20 by 9.
6. Multiply 24 by 7; 33 by 6; 87 by 2.
7. How many are 7 times 36?

NOTE. 7 times 3 tens are 21 tens, and 7 times 6 are 4 tens and 2; 21 tens are 2 hundreds and 1 ten.

8. How many are 9 times 38? 7 times 76? 4 times 57?

---

* MULTIPLICATION is the process of taking one number as many times as there are units in another number. The number to be multiplied, or taken, is called the *multiplicand*, the number by which we multiply is called the *multiplier*, and the result is the *product*.

9. Multiply 37 by 9; 49 by 7; 29 by 8.

10. How many are 8 times 18? 9 times 19? 9 times 28?

11. IIow many are 8 times 38? 6 times 49? 7 times 55?

12. How many are 4 times 48? 3 times 36?

13. Multiply 85 by 2; 58 by 3; 44 by 4; 35 by 5.

14. Multiply 29 by 6; 24 by 7; 29 by 9.

15. Multiply 39 by 9; 44 by 8; 51 by 7.

16. Multiply 61 by 6; 74 by 5; 93 by 4.

17. Multiply 18 by 14.

NOTE. 14 is 1 ten and 4; 10 times 18 is 180, and 4 times 18 is 72; 18 tens and 7 tens and 2 are 25 tens and 2, or 252.

18. Multiply 15 by 16; 15 by 12; 14 by 13.

19. Multiply 26 by 16; 23 by 13; 28 by 18.

20. How many are 13 times 17? 24 times 16? 15 times 15?

21. Required the product of 31 and 11; of 32 and 12; of 33 and 13.

22. Find the product of 18 and 19; of 19 and 20; of 20 and 21.

23. Multiply together 23 and 17; 22 and 18; 21 and 19.

24. Multiply together 28 and 16; 27 and 19; 26 and 20.

25. Find the product of 37 and 11; of 36 and 12; of 35 and 13.

26. How many are 11 times 44? 12 times 35? 13 times 32?

27. Multiply 11 by 18; 21 by 18; 31 by 18.

28. Multiply 12 by 17; 22 by 17; 32 by 17.

29. Multiply 13 by 14; 23 by 14; 33 by 14; 43 by 14.

30. Required the product of 17 by 17; of 27 by 17; of 37 by 17.

31. Required the product of 18 and 17; of 28 and 17; of 38 and 17.

32. Multiply 11 by 11; by 21; by 41.

33. Multiply 16 by 12; 18 by 13; 16 by 14; 28 by 14.

34. How many are 6 times 16? 16 times 16? 16 times 17? 16 times 18? 16 times 23? 18 times 25?

LESSON LVI.

1. Multiply 29 by 27.

NOTE. 29 by 2 tens and 7 are 58 tens and 203, or 20 tens and 3.

2. Multiply 28 by 28; 30 by 20; 29 by 21; 27 by 27.

3. How many are 23 times 27? 24 times 26? 25 times 25?

4. Multiply 27 and 18 together; 25 and 28; 24 and 29.

5. Multiply 26 and 23 together; 25 and 24.

6. Required the product of 24 and 32; of 24 and 34; of 25 and 35.

7. Find the product of 38 and 19; of 37 and 13; of 39 and 13.

8. How many are 14 times 17? 28 times 34?

9. Multiply 38 by 24; 37 by 25; 36 by 26; 35 by 27.

10. Multiply 26 by 36; 27 by 37; 28 by 38.

11. How many are 28 times 11? 28 times 21? 28 times 31?

12. How many are 26 times 14? 26 times 24? 16 times 24?

13. How many are 34 times 23? 36 times 22? 26 times 12? 26 times 22? 26 times 32?

14. Multiply 24 by 19.

Note. — It is easier to multiply 24 by 20, and take one 24 from the product, than to multiply the 24 by 10, and add 9 times 24. 24 × 20 = 480, and 480 — 24 = 456, the answer.

15. Multiply 34 by 19; 44 by 19.

16. How many are 19 times 32? 29 times 22? 39 times 12?

17. How many are 18 times 44?

Note. — A number can be multiplied by 18, 28, &c., by first multiplying by 2 more than the given multiplier, and then subtracting 2 times the multiplicand from the result obtained.

18. How many are 17 times 45? 16 times 46?

19. Multiply 37 by 35.

Note. 30 times 37 are 111 tens, or 11 hundreds and 1 ten, 1 thousand 1 hundred and 1 ten.

20. Multiply 47 by 23; 47 by 33; 57 by 33.

. 21. Multiply 33 by 37; 34 by 36; 35 by 35.

22. How many are 35 times 33? 36 times 36?

23. Required the product of 44 by 44; of 43 by 43; of 42 by 42.

24. Multiply 37 by 29, by subtraction.

Note. — That is, subtract once 37 from 30 times 37.

25. Multiply 28 by 39; 39 by 39; 38 by 38; 37 by 37.

26. How many are 29 times 29? 29 times 28? 29 times 27?

27. How many are 39 times 47? 49 times 37? 59 times 27? 59 times 28? 59 times 29?

28. Multiply 50 by 24; 49 by 25; 48 by 26; 47 by 29; 46 by 28.

29. Multiply 40 by 34; 39 by 35; 38 by 36; 39 by 37.

30. Multiply 67 by 63.

Note. 67 by 60 is 4020; 67 by 3 is 201; and 4020 and 201 is 4221.

31. Multiply 61 by 69; 62 by 68; 67 by 67.

32. Required the product of 64 by 83; 63 by 84; 55 by 57.

33. Multiply 71 by 39; by 49; by 59.

34. Multiply 70 by 70; 39 by 40; 83 by 98.

35. Find the product of 59 and 61; of 59 and 62; of 42 by 71.

36. Multiply 48 by 48; 58 by 58; 95 by 97; 93 by 95; 80 by 96.

LESSON LVII.

1. Find the square of 11.

Note. — The square of any number is the product of that number by itself. The square of 11 is 11 times 11, or 10 times 11 and once 11; 10 times 11 is 10 times 10 and once 10; 11 times 11 is 10 times 10, and 10 and 11, or 121.

2. How much greater is the square of 12 than the square of 11?

3. How much greater is the square of 13 than the square of 11?

4. If two numbers, whose difference is 1, have a sum of 27, what is the difference of their squares?

Note. — The greater number multiplied by itself is equal to the greater multiplied by the less, plus the greater; the greater multiplied by the less is the less by the less, plus the less; the greater multiplied by itself is equal to the less multiplied by itself, plus the greater, plus the less; that is, the square of the greater is equal to the square of the less, plus their sum.

5. If the sum of two numbers is 35, and their difference 1, what is the difference of their squares?

6. If the sum of two numbers is 18, and their difference 2, what is the difference of their squares?

Note. — Their difference being 2, the difference of their squares is 2 times their sum. For the greater multiplied by the less is 2 times the less, plus the square of the less; and the square of the greater is 2 times the greater, plus the product of the two.

7. If the sum of two numbers is 24, and their difference 2, what is the difference of their squares?

8. If the sum of two numbers is 27, and their difference 3, what is the difference of their squares?

9. What is the sum of 13 and 10? What is their difference? What is the difference of their squares? What is the square of 10? Of 13?

10. What is the square of 15?

NOTE.—Find the square of 10, then the difference of the squares of 10 and 15, then the square of 15.

11. What is the square of 20? What is the difference of the squares of 21 and 20? What is the square of 21?

12. What is the square of 23?

NOTE.—To find the square of any number which has two figures, find the square of the next less even number of tens, 2 tens, or 20, in this case; then find the difference of the square of their tens, and the given number; the square of $20 = 400$; 3 times $20 + 23 = 129$; $400 + 129 = 529$, the answer.

13. What is the square of 25? Of 27?
14. What is the square of 30? Of 31?
15. What is the square of 33? Of 35? Of 37?
16. What is the square of 52? Of 55? Of 58?
17. What is the square of 61? Of 65? Of 73?
18. What is the square of 75? Of 81? Of 87?
19. What is the square of 102? Of 110? Of 115?
20. What is the square of 120? Of 127? Of 133?
21. What is the square of 139? Of 145? Of 151?
22. What is the square of 189? Of 191? Of 187?
23. What is the square of 215? Of 217? Of 229?

Note. — The square of 229 can also be obtained by subtracting the difference of the squares of 229 and 280 from the square of 280.

24. What is the square of 235? Of 245? Of 255?

25. What is the square of 292? Of 333? Of 365?

26. What is the square of 315? Of 325? Of 282? Of 184?

27. What is the square of 254? Of 262? Of 281?

28. What is the square of 303? Of 311? Of 319? Of 322?

29. What is the square of 356? Of 357? Of 325?

30. What is the square of 5½?

Note. — When the fractional part of a mixed number is ½, a convenient way of finding the square is to multiply the whole number of the given expression by a number 1 larger than the given whole number, and to the product add the square of the fractional part. Thus, $5 \times 6 = 30$; $\frac{1}{2} \times \frac{1}{2} = \frac{1}{4}$; and $30 + \frac{1}{4} = 30\frac{1}{4}$.

31. Multiply 4½ by 4½; 3½ by 3½.
32. Multiply 7½ by 7½; 10½ by 10½.
33. Multiply 20½ by 20½; 30½ by 30½.
34. Multiply 9 by 11.

Note. 11 times 9 is 11 times 10, less once 11; but 11 times 10 is 10 times 10, plus once 10; then 11 times 9 is 10 times 10, less 1, for 10 is 11, less 1.

35. Multiply 12 by 8; 13 by 7; 14 by 6.

Note. — The product of the sum of two numbers by their difference is the difference of their squares.

36. Of what two numbers is 13 the sum, and 7 the difference?

Note. — The sum of two numbers, plus their difference, is twice the greater; the sum, less the difference, is twice the less.

37.   What is the product of 14 and 6 ?

Note. — Since 14 is the sum of 10 and 4, and 6 their difference, the product of 14 and 6 is the difference of the squares of 10 and 4, or 100 — 16 = 84.

38.   What is the product of 19 and 21 ?
39.   How many are 29 times 31 ?   28 times 32 ? 27 times 33 ?
40.   How many are 34 times 46 ?   36 times 44 ? 35 times 45 ?
41.   Multiply 355 by 113.*

Note.   100 times 355 is 100 times 300, plus 100 times 55 ; 13 times 355 is 13 times 300, plus 13 times 55.   Then 113 times 355 is 100 times 300, plus 100 times 55 and 13 times 300, plus 13 times 55.   100 times 300 is 30000 ; 100 times 55 is 5500, and 13 times 300 is 3900 ; 13 times 55 is 715.   Now, the first number is expressed in *ten thousands ;* the second and third in *hundreds,* and the fourth in units.   Add the two sets of hundreds together, and to them add the hundreds from the product of units, for the hundreds : if there are more than a hundred of hundreds, add the number of hundred of hundreds to the ten thousands.   Thus, 715 equals 15 and 700 ; and 700, 3900 and 5500 are 101 hundreds ; 100 hundreds equal 10000, which, added to 30000 gives 40000 ; and 15, 100 and 40000 equal 40115.

42.   Multiply 121 by 114.

Note.   21 by 14 is 294 ; and 294 equals 94 and 200 ; add the 200 to the hundreds, which are 100 by 14, and 100 by 21, or 14 hundreds and 21 hundreds ; the sum of the hundreds is 3700 ; 100 times 100 is 10000, which, added to 8700 and 94, gives 13794.

43.   What cost 218 horses, at $142 apiece ?
44.   Multiply 136 by 137 ; 128 by 164 : 121 by 137.
45.   What is the square of 333 ?

---

* The remaining exercises in this lesson, if thought to be too difficult for some classes, at the option of the teacher, may be omitted till a review.

46. How many square inches in 162 square feet ?
47. Multiply 1234 by 1326.

NOTE. $1200 \times 1826 = 1200 \times 1300 + 1200 \times 26$;—and $84 \times 1326 = 84 \times 1800 + 84 \cdot \times 26$; and $84 \times 26 = 884$. Then the last two figures will be 84 ; there will be 8 hundreds to add to the hundreds. $1200 \times 26 = 312$ hundreds ; $84 \times 1800 = 442$ hundreds ; 8, 312 and 442 are 762. The two figures just before the last two of the result sought will be 62, and there will be 7 hundreds of hundreds to add to 1200×1300 ; $1200 \times 1300 = 1560000$; 7 and 156 are 163. The whole product sought will therefore be 163 hundreds of hundreds, or tens of thousands, and 62 hundreds, and 84 units, or 1636284 (read 1 million, 636 thousand, 2 hundred 84.)

48. Multiply 1213 by 1156 ; 1679 by 655.
49. Find the square of 1333.

NOTE. $33 \times 33 = 1089$; 33×1300 is to be taken twice, which gives 66×1800, or 85800 ; $1300 \times 1300 = 1690000$; $1690000 + 85800 + 1089 = 1776889$.

LESSON LVIII.

1. Divide 88 by 8.*

NOTE. 88 equals 8 tens and 8 units ; 8 in 8 tens 1 ten time, and 8 in 8 units 1 unit time ; hence, 1 ten and 1 unit, or 11, is the answer.

2. Divide 90 by 6.

NOTE. 6 in 9 tens 1 ten time, and 3 tens remainder ; 6 in 3 tens, or 80, 5 times ; answer, 15.

3. Divide 96 by 8.

---

* DIVISION is the process of finding how many times one number is contained in another, or the process of separating a number into any proposed number of equal parts. The number to be divided is called the *dividend ;* the number by which we divide is called the *divisor ;* and the result, or number produced by the division, is called the *quotient ;* and the excess that is sometimes left after dividing, is called the *remainder.*

NOTE. 8 in 9 tens and 6 units 1 ten time, and 1 ten and 6 units remainder ; 8 in 1 ten and 6 units, or 16, 2 times ; answer, 12.

4. Divide 50 by 5 ; 60 by 5 ; 75 by 5 ; 85 by 5 ; 90 by 5.

5. Divide 66 by 6 ; 78 by 6 ; 84 by 6.

6. Divide 84 by 7 ; 98 by 7 ; 77 by 7.

7. Divide 99 by 9 ; 63 by 7 ; 88 by 4.

8. 126 contains 6 how many times ?

NOTE. 126 equals 12 tens and 6 units.

9. How many times will 126 contain 7 ? Divide 133 by 7 ; 184 by 8 ; 224 by 8.

10. Divide 252 by 3 ; by 4 ; by 6 ; by 7 ; by 9.

11. How many quarts of milk, at 6 cents a quart, can you buy for 9 dimes and 6 cents ? How many for 2 dollars, 2 dimes, 2 cents ?

12. If 1 dozen bottles of beer cost 96 cents, how much costs 1 bottle ? How much cost 26 bottles ? How many bottles can be bought for $ 3.12 ?

13. Find the quotient of 336 divided by 8.

14. Divide 648 by 9 ; 666 by 9 ; 567 by 9.

15. Divide 357 by 7 ; 434 by 7 ; 656 by 8.

16. Divide 736 by 8 ; 648 by 8 ; 568 by 8.

17. Divide 129 by 8 ; 172 by 9 ; 181 by 9.

NOTE. — Here we have a remainder.

18. Divide 169 by 7 ; 168 by 7 ; 168 by 8.

19. How many times is 7 contained in 169 ? In 172 ? In 175 ?

20. How many barrels of flour, at $9 a barrel, can be bought for $162 ?

21. Divide 868 by 8 ; 808 by 8 ; 848 by 8 ; 888 by 8.

NOTE. 80 tens divided by 8, gives 10 tens, or 1 hundred.

22. Divide 896 by 7.

Note. — We find the *hundreds* in the quotient first ; 7 is contained in 8 hundreds 1 hundred times, and 1 hundred over.

23.   Divide 968 by 8 ; 1048 by 8 ; 1136 by 8.

24.   How many times is 8 contained in 968 ?   In 1048 ?   In 1136 ?

25.   How many times is 5 contained in 644 ?   In 1425 ?

26.   Divide 1448 by 8 ; 1656 by 8.

27.   What is the quotient of 867 divided by 9 ?

28.   How many times is 8 contained in 1843 ? In 1544 ?

29.   How many hours will it take a steamer to sail 999 miles, at 9 miles an hour ?

30.   If $1000 be divided among 9 men, how much will each man receive ?

31.   How many times is 9 contained in 10 ?   In 100 ?

32.   How much is left when you divide 176 by 9 ?

Note.   1 hundred divided by 9 leaves 1 remainder ; 7 tens divided by 9 leaves 7 remainder ; 6 units divided by 9 leaves 6 remainder.   Therefore, 176 divided by 9 leaves 1 and 7 and 6 remainder ; but 1 and 7 and 6 are 14, which, divided by 9, leaves 5 remainder.

Every number divided by 9, leaves just as much remainder as all its figures, taken as units and added together, would.   For *every* unit, ten, hundred, or thousand, leaves 1 remainder when divided by 9.

33.   How much is left after dividing 172 by 9 ? 876 by 9 ?   929 by 9 ?

34.   How many pounds of sugar can be bought for $14.24, at 8 cents a pound ?

35.   At 7 cents apiece, how many slates can be bought for $16.38 ?

36.   How many pecks in 3853 quarts ?   In 6353 quarts ?

37.   If 9 yards are required for 1 cloak, how many cloaks may be made from 6354 yards ?

LESSON LIX.

1. Divide 121 by 11.

NOTE. — We first seek how many figures on the left of 121 will contain 11 more than once, and less than 10 times : 11 will go into 1 not at all ; into 12 it will go once, and a remainder. Then, as there is *one* figure left after 12, in the dividend, the quotient will have two figures. 11 goes into 12 tens, 1 ten time, and 1 ten over ; 1 ten and 1 unit are 11, into which 11 goes 1 time ; 1 ten and 1 unit, or 11, is the answer.

2. Divide 132 by 11 ; 143 by 11.
3. Divide 144 by 12 ; 168 by 12 ; 180 by 12.
4. Divide 221 by 13 ; 216 by 18 ; 225 by 15.
5. What number is that which will be contained 16 times in 256 ?
6. What number is that which will be contained 18 times in 270 ?
7. How many school libraries of 230 volumes may be selected from 1840 volumes ?
8. How many companies of 80 soldiers each may be formed from 1920 soldiers ?
9. How many cases of 48 pieces each can be put up from 2400 pieces ?
10. At 16 cents a yard, how many yards of calico can be bought for $13.12 ?
11. Divide the product of 74 and 63 by 18.
12. How many times will 16 go into 512 ?
13. What number will go into 512 16 times ?
14. If the dividend is 666, and the divisor 18, what is the quotient ?
15. If the dividend is 666, and the quotient 18, what is the divisor ?
16. If the divisor is 21, and the quotient 23, what is the dividend ?
17. If the quotient is 21, and the divisor 23, what is the dividend ?

18. If the multiplicand is 27, and the multiplier 23, what is the product?

19. Divide 621 by 23; 621 by 27.

20. If the product is 621, and the multiplier 23, what is the multiplicand?

21. If the product is 621, and the multiplicand 27, what is the multiplier?

22. Divide 672 by 24; 792 by 24; 912 by 24.

23. Divide 784 by 28; 868 by 28; 980 by 28.

24. Divide 1001 by 11; 1920 by 24; 2304 by 24.·

25. How many times will 37 go into 1369? 45 into 2160? 31 into 992?

26. If the dividend is 1216, and the divisor 44, what is the quotient?

27. How, many times will 39 go into 1209? Into 1755? Into 2496?

28. Find the number of times 41 is contained in 2829; in 2501; in 3321.

29. Required the number of times 47 is contained in 2444; in 3666.

30. Find how many times 59 will go into 4012; into 4661.

NOTE. 59 will go 80 times into 4720; 4661 is how much less than 4720?

31. If a train of cars move at the rate of 34 miles an hour, how many hours will it be in moving 1187 miles? 1189 miles?

32. How many days will 3484 gallons of water supply a family, if 67 gallons of it are used daily? 3488 gallons? 5025 gallons?

33. Divide 4212 by 78; 5109 by 78; 6097 by 78.

NOTE. — In any case, when it can be readily done, reduce the fraction formed by writing the divisor under the remainder, to its lowest terms.

34. If 94 bushels of corn can be produced on 1 acre, at the same rate, how many acres will be required to produce 5264 bushels? 5292 bushels?

LESSON LX.

1. Find the factors of 32.*

NOTE. — Factors of a number are such numbers as will, by being multiplied together, produce that number. Unity, or 1, is not regarded as a material factor, since multiplying or dividing any number by 1 does not alter its value. It will be disregarded, therefore, when speaking of the factors of numbers. 2 is a factor of any even number. 2 and 16 are factors of 32.

2. Find the factors of 16; of 8; of 4; of 2.

NOTE. — When a number has no factor but itself and unity, it is called a *prime number*. 2 is therefore a prime number. Factors that are prime numbers are called prime factors. The factors of 32, if we take account of them all, are either 5 *twos*, or 3 *twos* and 1 *four*, or 2 *twos* and 1 *eight*, or 1 *two* and 1 *sixteen*, or 1 *two* and 2 *fours*.

3. Find the prime factors of 12; of 6; of 3.

NOTE. 3, having no factors but itself and 1, is a prime number.

4. What prime numbers are there under 10?
5. Required the prime factors of 14.
6. Which are the two factors of 56 which are nearest equal? ·

NOTE. 56 = 28 × 2, or 14 × 4, or 8 × 7.

7. Divide 63 into 3 factors.
8. What are the prime factors of 49? Of 77? Of 91?

---

* The process of resolving a quantity into its factors is called *factoring*.

9. How many prime numbers are there among the twenties?

NOTE. — No even number is prime, for all even numbers can be divided exactly by 2. All whole numbers whose last figure is even, are even. No whole number whose last figure ends in 5 is prime, for all such numbers can be divided by 5. Any number of *tens* can be divided by 2 or 5.

10. Required the factors of 35; of 39.

11. How many prime numbers among the thirties?

12. How many times does 7 occur as a factor in 98?

13. I have 42 bushels of grain; which are the three smallest sizes of bags, holding an exact whole number of bushels, that will exactly contain the same? How many bags of the smallest size would exactly contain the grain? How many of each of the other two sizes?

14. Multiply 53 by 28.

NOTE. 28 is 7 times 4; 7 and 4 are the factors, therefore, of 28. To multiply 53 by 28, we may use the factors of the 28, multiplying first by 7, and then its product by 4.

In finding the factors of any number less than 100, we need divide only by 2, 3, 5 and 7, since, if it cannot be exactly divided by any of these, it is prime.

15. Multiply 49 by 45, using the factors of 45.

NOTE. 45 can be divided by 5, because its last figure is 5. Its other factor is 9.

16. Multiply 47 by 54.

NOTE. — All questions of this kind may be solved by using factors, or by the use of *tens* and *units*.

17. What cost 62 cows, at $39 apiece?

18. At $84 an acre, what cost 59 acres of land? 73 acres? 14 acres?

19. Find the product of 18 by 32; 48 by 42; 68 by 64.

20. Multiply 127 by 121, using factors.

NOTE. — To find the factors of a number less than 400, we may divide by any prime number not over 20 ; that is by 2, 8, 5, 7, 11, 13, 17 and 19. If the number is more than 400, and less than 900, we can also divide by 23 and 29 ; and if between 900 and 1000, also by 81.

21. What are the factors of 127 ? Of 143 ? Of 221 ? Of 323 ? Of 483 ?

22. What are the factors of 162 ? Of 328 ? Of 492 ? Of 564 ?

NOTE. — In the last number we find 4×141 ; we then can find the factors of 141, and so on.

23. Multiply 195 by 132 ; 112 by 154 ; 311 by 686, by using the factors.

24. Multiply by 10 the numbers 13 ; 123 ; 265 and 884.

NOTE. — We may use the factors in multiplying by 10, 100, and so on, or we may annex a 0, or cipher, to the multiplicand, to multiply it by 10 ; two ciphers to multiply it by 100 ; and so on.

25. Multiply by 10 the numbers 924 ; 572 ; 763 ; 292 and 29.

26. How many dimes in $ 10 ? In $ 129 ? In $ 355 ? In $ 762 ?

27. Multiply 253 by 10 ; by 100 ; by 1000.

28. Multiply 303 by 6 ; by 60 ; by 600.

29. How many cents in $ 5 ? In $ 50 ? In $ 72 ? In $ 125 ?

30. How many dollars in 265 eagles ? How many dimes ? How many cents ? How many mills ?

31. What cost 218 yards of cloth, at $ 1.30 a yard ?

32. How many pounds in 56 tons, at 2000 pounds to a ton ?

33. Divide 1200 by 10 ; by 100.

NOTE. — We may divide by 10, by cutting off a figure from the right of the dividend ; by 100, by cutting off two figures, and so on ; the figure or figures cut off, if other than a cipher or ciphers, must be considered as a remainder.

34. Divide 1000 by 10 ; 1500 by 10 ; 1700 by 100.

35. How many dollars are there in 30 dimes? In 300 dimes? In 900 dimes?

36. How many dollars in 500 cents? In 3000 mills?

37. At $20 an acre, how many acres of land can be bought for $1240?

38. At $300 each, how many horses can be bought for $5400?

39. How many times will 130 go into 11800?

40. How many village lots, at $159 each, may be bought for $10494?

NOTE. — Resolve 159 into its prime factors, and divide by the factors in succession.

LESSON LXI.

1. What kind of a fraction is $\frac{8}{10}$?

NOTE. — A fraction whose denominator is 10, 100, 1000, and so on, is called a *decimal* fraction ; other fractions are called *common* fractions.

2. What kind of a fraction is $\frac{1}{4}$? How can you make a decimal fraction of it?

NOTE. — When the denominator of any fraction is changed to 10, 100, 1000, or so on, the fraction becomes a decimal.

3. Change $\frac{14}{25}$ to a decimal fraction.

NOTE. — By multiplying both terms of the fraction by 4 which does not change its value, we have $\frac{56}{100}$, the answer required.

4. Change $\frac{7}{50}$ to a decimal fraction.

NOTE. — To change 30ths to 10ths, find how many 10ths there are in as many whole ones as you have 30ths, and then divide them by 30.

5. Change $\frac{27}{45}$ to a decimal fraction.
6. Change $\frac{33}{44}$ to a decimal fraction.

NOTE. — First reduce the fraction to its lowest terms. Find then what product of 10 by itself can be divided by the denom. inator.

7. What decimal fraction is equal to $\frac{4}{5}$? To $\frac{8}{20}$? To $\frac{28}{40}$?
8. What does 0.7 signify?

NOTE. — The point after the 0 is called the *decimal point*; all the figures before it denote whole numbers; all after it decimal fractions.

If there is but one figure after the decimal point, it is *tenths*; if two, they are *hundredths*; if three, *thousandths*; and so on. 0.6 is $\frac{6}{10}$; 0.64 is $\frac{64}{100}$; 0.648 is $\frac{648}{1000}$; 1.5 is $1\frac{5}{10}$; 22.4 is $22\frac{4}{10}$; 365.252 is $365\frac{252}{1000}$.

9. What does 1.8 signify? 2.81? 12.892?
10. What does 6.872 signify? 0.97? 5.5?
11. What does $124.54 signify?

NOTE. — Cents are decimals of a dollar, so dollars and cents are written with the decimal point between them.

12. What does $223.6 signify? $223.06? $223.60?

NOTE. — We do not often use the decimal expression for dimes, but rather annex a cipher and make cents of dimes, just as we talk of ten *cents* instead of one *dime*.

13. What do the figures 0.24 express? The figures 0.46? The figures 3.1? The figures 3.14?
14. What part of a dollar is 48 cents? 34 cents? 34 cents 3 mills?
15. Reduce 1 shilling to the decimal of a pound; 3 shillings; 7 shillings.
16. Reduce 7 shillings 6 pence to shillings and tenths; to the decimal of a pound.

17.   Reduce $\frac{1}{3}$ to a decimal.

NOTE. $\frac{1}{3}$ is equal to no even number of tenths, but it is nearly equal to 0.3 ; nearer to 0.33 ; still nearer to 0.333 ; and so on. If we have $\frac{1}{3}$ of a dime to reduce to cents, we should call it 3 cents ; $\frac{1}{3}$ of a dollar, we should call it 33 cents ; and $\frac{1}{3}$ of an eagle $2.33 ; and so on.

18.   How many cents in $\frac{1}{5}$ of a dime ?

NOTE. $\frac{1}{5}$ of 10 cents is nearly 2 cents.

19.   How many cents in $\frac{1}{7}$ of a dollar ? $\frac{1}{8}$? $\frac{1}{9}$?
20.   Reduce $\frac{3}{7}$ of an eagle to dollars and cents.
21.   What decimal to *hundredths* of a pound avoirdupois is 1 ounce ? Is 3 ounces ? Is 4 ounces ?

NOTE. — " *To hundredths*," means that all fractions less than hundredths are not counted — more than 5 thousandths counting as an additional hundredth.

22.   How many hundredths of a foot are there in 6 inches ? In $7\frac{1}{2}$ inches ? In $8\frac{1}{4}$ inches ?
23.   How many hundredths of a mile are there in 154 rods ? How many thousandths of a mile in 129 rods ?

NOTE. 320 rods make a mile.

24.   Reduce 67 square inches to the decimal of a foot ; 79 square inches ; 87 square inches.

NOTE. — If inconvenient to divide by 144, divide by 12, twice, since 12 times 12 are 144.

25.   Reduce 85 square rods to thousandths of an acre ; $97\frac{1}{2}$ square rods ?
26.   Reduce 85 cubic feet to the decimal of a cord, to hundredths ; 77 cubic feet ; 44 cubic feet.
27.   Reduce 22 hours to the decimal of a day ; 16 hours 45 minutes ; 15 hours 23 minutes.

NOTE. — There are 1440 minutes in a day.

28.   Reduce 12 ounces to the decimal of a pound ; 8 ounces ; $10\frac{1}{2}$ ounces.

LESSON LXII.

1. Add together 0.25, 0.67 and 0.92.
2. Add together $1.26, $1.37 and $0.63.
3. How much are $22.54 and $27.46?
4. Add together 0.9, 0.09, 0.009 and 0.27.
5. How much are $2.65, $3.1 and 1 cent?
6. Add $\frac{4}{5}$ to 2.76; $\frac{2}{5}$ to 2.21; $\frac{13}{20}$ to 4.35; $\frac{27}{50}$ to 4.04.
7. Subtract 0.645 from 3.334.
8. How much is $77.25 less $64.51?
9. Subtract 0.62 from $5\frac{3}{4}$.
10. Subtract 2.228 from $3\frac{1}{8}$.
11. Subtract 1.58 from $1\frac{7}{12}$, to thousandths.
12. What is the sum of 3.12, $5\frac{1}{4}$ and 6.38, less, $2\frac{1}{8}$?
13. From $3\frac{3}{4}$ added to 2.72, subtract $2\frac{1}{5}$ added to $3\frac{1}{4}$.
14. Multiply 24 by 0.008; 45 by 0.9.
15. Multiply 1.27 by 11; 2.35 by 8; 3.27 by 9.
16. What is the product of 346 by 0.02; 154 by 1.3?
17. What will 28 tons of coal cost, at $7.80 a ton?
18. Multiply 0.1 by 0.1; 0.1 by 0.01; 0.01 by 0.01.

NOTE. — When the multiplicand has one decimal place, and the multiplier one, how many will the product have? When the multiplicand has one decimal place, and the multiplier two, how many will the product have? When the multiplicand has two decimal places, and the multiplier two, how many will the product have?

19. Multiply 0.0009 by 10; by 100; by 1000.

NOTE. — Multiplying any decimal by 10 is equivalent to removing the decimal point one place nearer the end; and multiplying by 100 removes the decimal point two places nearer the end; and so on.

20. Multiply 0.007 by 10 ; by 100 ; by 1000.

21. What is the profit on 900 yards of cotton cloth, at $0.009 a yard?

22. What cost 10 barrels of flour, at $8.375 a barrel? 100 barrels? 1000 barrels?

23. Multiply $959 by 0.01 ; by 0.06 ; by 0.12.

24. How much is 1 per cent. of $959? 6 per cent.? 12 per cent.?

25. How much is 3 per cent. of $150? $\frac{3}{10}$ per cent.?

26. How much is $3\frac{3}{10}$ per cent. of $240?

27. How much is $5\frac{8}{10}$ per cent. of $262?

NOTE. $\frac{8}{10}$ per cent. of $262 is $2.096 ; as 0.006 is *more* than ½ of a cent, 0.096 is *nearer* 10 cents than 9 cents. $2.10 is sufficiently exact for $\frac{8}{10}$ per cent. of $262.

28. What is $8\frac{7}{10}$ per cent. of $536?

29. What is 6 per cent. of $424? 18 per cent.?

30. What is 7 per cent. of $216.40? 28 per cent.?

31. Required 33 per cent. of $200. 66 per cent.

NOTE. 33 per cent. is ⅓ of 99 per cent

32. If 100 tons of hay cost $20.27, how much costs 1 ton? How much cost 2.5 tons, leaving out fractions of a cent?

NOTE. —In leaving out fractions of a cent, call everything over 5 mills, 1 cent.

33. Required $16\frac{27}{100}$ per cent. of $212.54.

NOTE. 16 per cent. of $200 is $32 ; of $12 is $1.92 ; of $0.54 is $0.09 ; and $82 + $1.92 + $0.09 = $34.01, or 16 per cent. of $212.54. $\frac{27}{100}$ per cent. of $200 is $0.54 ; of $12 is $0.03 ; of $0.54 is $0.00, less than ½ of a cent ; and $0.54 + $0.03 = $0.57, or $\frac{27}{100}$ per cent. of $212.54. $34.01 + $0.57 = $34.58, the answer.

34. What is $8\frac{4}{10}$ per cent. of $67.33? $18\frac{7}{10}$ per cent. of $243.50?

35. What is the value of $306\frac{4}{10}$ pounds of tea, at $58\frac{4}{10}$ cents a pound?

NOTE. — At $1.00 a pound, it would be worth $306.40 ; so we must find $58\frac{4}{10}$ per cent. of $306.40.

36. What cost $8\frac{84}{100}$ cords of wood, at $6.43 a cord?

37. At $1.62 a day, how much can a man earn in 8.74 days?

38. How much will 262 pounds of beef cost, at $7.83 per hundred pounds?

LESSON LXIII.

1. Divide 8.96 by 7 ; 8.54 by 7 ; 9.80 by 8.

NOTE. 8 divided by 7 gives 1, and 1 over ; .96 and 1 are $\frac{196}{100}$; $\frac{196}{100}$ divided by 7 gives $\frac{28}{100}$, or .28.

2. Divide 2.60 by 4 ; 3.4 by 8 ; 5.4 by 12.
3. Divide 21 by 0.7 ; by 0.07 ; by 0.007.
4. Divide 28.4 by 16 ; 63.24 by 15 ; 48.45 by 11, to hundredths.
5. Divide 36.3 by 9, to thousandths.
6. How many times is 16 contained in 256 ? In 2560 ?
7. How many times is 1.6 contained in 25.6 ? In 256 ?
8. How many times is 7 contained in 63.7 ? In 6.37 ?
9. Divide 1 to thousandths, by 6 ; by 7 ; by 8 ; by 9.
10. How many times is 1.4 contained in 2.24 ?

NOTE. $\frac{14}{10}$ is contained in $\frac{224}{10}$ 16 times ; how many times in $\frac{1}{10}$ of $\frac{224}{10}$?

11. Divide 1 by 3, to hundredths; to thousandths; to ten thousandths.

12. Divide 16.9 by 1.3; 1.69 by 1.3; 1.69 by 0.13.

13. Divide, to hundredths, 1.82 by 0.12; 18.2 by 1.2; 182 by 12.

14. Divide $164.2 equally among 7 persons, without counting fractions less than $\frac{1}{2}$ of a cent.

15. What is the value of $\frac{1}{3}$, to thousandths? Of $\frac{1}{6}$, to thousandths?

16. How many dollars, dimes, and cents in $\frac{1}{4}$ of an eagle? How many dollars, cents and mills?

NOTE. — By cents are meant *whole* cents; that is, fractions less than $\frac{1}{2}$ of a cent are to be neglected, and fractions from $\frac{1}{2}$ a cent to 1 cent are to be called 1 cent.

17. How many dollars and cents in $\frac{1}{6}$ of an eagle? How many dimes, cents and mills?

18. $\frac{1}{11}$ of 1 is how many thousandths, nearly?

19. $\frac{1}{13}$ of 1 is how many hundredths, nearly?

20. Divide 2 by 0.01; 3 by 0.001; 1 by 0.08; 1 by 0.008.

21. How many pounds of sugar, at 11 cents a pound, can be bought for $1.76?

22. On what principal will the interest for 0.7 years, at 6 per cent., be $2.52?

NOTE. — First find how much the interest on $1 for 0.7 years will be.

23. How many yards of cloth, at $26\frac{4}{10}$ cents a yard, can be bought for $1.77?

24. Divide 1.77 by 0.3; by 0.003; by 3.

25. Divide 18.3 by 7, to hundredths; by 0.7, to hundredths; by 0.07, to tenths.

26. How long will it take a man to walk 26.45 miles, at the rate of 5 miles an hour?

27. At $\frac{6}{10}$ of a dollar a bushel, how many bushels of corn can be bought for $7.20?

28. When corn is $\frac{5}{10}$ of a dollar, and rye $\frac{2}{10}$ of a dollar a bushel, how many bushels, consisting of corn and rye in equal parts, can be bought for $7.35?

LESSON LXIV.

1. If 12 yards of cotton cloth cost $1.60, how much will 39 yards cost?

NOTE. 39 is 3 times 12, and 3 over; 3 is $\frac{1}{4}$ of 12. Then 3 times the cost of 12 yards gives $4.80, and $\frac{1}{4}$ the cost of 12 yards, $0.40.

2. If 7 hundred-weight of beef cost $64, how much will 23 hundred-weight cost?

NOTE. 23 hundred-weight is 3 times 7 hundred-weight, and 2 hundred-weight over, and 2 hundred-weight is 1 quarter of 7 hundred-weight, and 1 quarter of 1 hundred-weight.

3. If 5 thousand of bricks cost $30, how much will 28 thousand cost?

4. 11 is to 33 as $0.54 is to what sum?

5. If 3 and 4 were 12, what, on the same supposition, would 2 and 3$\frac{1}{4}$ be?

6. If the difference of two numbers is 29, and the greater is 207, what is the less?

7. When the difference between two numbers, and the greater number are given, how do you find the less?

8. If the difference of two numbers is 28, and the less 174, what is the greater?

9. When the difference between two numbers, and the less number are given, how do you find the greater?

10. Bought 232 pounds of beef, at 11 cents a pound, and sold half of it for 13 cents, and half for 14 cents a pound; how much did I gain?

11. Bought 28 yards of broadcloth, at $6 per

13

yard, and. paid for the same in wood, at $7 per cord ; how much did it take ?

12. What cost 96 bushels of potatoes, at 25 cents a bushel ?

NOTE. 25 cents is ¼ of a dollar.

13. If the sum of two numbers is 5, and their difference is 3, what are the numbers ?

NOTE. — The sum of two numbers, added to their difference, gives twice the greater of the numbers. Because, in adding the sum and difference, the greater is added to itself, and the less subtracted from itself. In subtracting the difference from the sum, we subtract from the greater increased by the less, itself decreased by the less. This gives twice the less; *once* the greater in the sum, and *once* the less in the difference. Hence, if the sum of two numbers is 5, and their difference 3, the greater is ½ of 5 and 3, or 4 ; and the less is ½ of 5 less 3, or 1.

14. The sum of two numbers is 17, and their difference 5 ; what are they ?

15. The sum of two numbers is 65, and their difference 19 ; what are they ?

16. Two boys, on counting their money, found that the one had $46 more than the other, and together they had $54 ; how much had they each ?

17. The sum of two numbers is 152, and their difference 68 ; what are they ?

18. If 25 barrels of flour cost $131¼, how much will 3½ barrels cost ?

19. What cost $\frac{7}{8}$ of a hogshead of molasses, at $\frac{3}{8}$ of a dollar a gallon ?

20. What number, increased by ¼, by ½, and by ⅕ of itself, will amount to 39 ?

21. A and B undertake to travel round a circular island 20 miles in circuit, both starting from the same point, and going round in the same direction ; when A has travelled 19 miles, and B 8, how far apart are they ?

22. Divide $100 between A, B and C, so that

B shall have $20 more than A, and C $15 less than B.

23. A man who has 60 sheep and lambs, finds that the number of the sheep, less the number of lambs, is $\frac{2}{5}$ of the flock; how many are there of each kind?

24. What part of the principal is $7\frac{1}{2}$ per cent. interest?

25. At what per cent. interest will $10 become $15, in 5 years?

26. The difference of two numbers is 5, and the less number is $\frac{4}{9}$ of the greater; what are the numbers?

27. When an article is sold at $\frac{2}{3}$ of its cost, what is the loss per cent.?

28. John Smith gave 9 cents a dozen for apples, and had 12 cents left; but, had he paid 12 cents a dozen, he would have spent all his money; how many apples did he buy?

NOTE. — The 12 cents left was how many cents a dozen?

29. A man bought a certain number of quires of paper, at 18 cents a quire, and had 84 cents left; had he bought just as many quires at 24 cents a quire, he would have had 6 cents left; how much money had he?

30. Bought rice at $6\frac{1}{2}$ cents a pound, and sold it at $8\frac{1}{8}$ cents a pound; what per cent. was the gain?

31. A gentleman bought 15 cords of wood, oak and pine, for $55; he paid $5 a cord for the oak, and $3 a cord for the pine; how many cords were there of each, and how much did each kind cost?

32. A laborer agreed to work 30 days for $2 a day, and for every day he was idle to pay $\frac{3}{4}$ of a dollar for his board; he received $38; how many days was he idle?

LESSON LXV.

1. How many yards of cambric which is $\frac{3}{4}$ of a yard wide, will be required to line 30 yards of cloth that is $1\frac{3}{4}$ yards wide?

2. In a certain school $\frac{1}{8}$ of the pupils study algebra, $\frac{1}{4}$ geometry, and the remainder arithmetic; what per cent. of the whole are in each of the studies named?

3. $\frac{1}{2}$ less $\frac{1}{4}$, plus $\frac{3}{4}$, multiplied by $1\frac{1}{3}$, is how many times $\frac{1}{4}$?

4. A dog one night killed 17 sheep, which was $5\frac{2}{3}$ per cent. of a flock; how many of the flock were spared?

5. If $4\frac{1}{5}$ bushels of wheat cost $\$9\frac{9}{20}$, what cost $\frac{3}{4}$ of a bushel?

6. Joseph earned $\$1\frac{1}{2}$ a day, and spent $\$3.50$ a week for board, and $\frac{1}{3}$ of the remainder of his earnings for clothes; at that rate, how much can he save in 4 weeks?

7. If two men together can mow a field in 10 days, and one of them alone can mow it in 15 days, in what time can the other mow it?

8. The head of a fish is 9 inches long; the tail is as long as the head and half the body, and the body is as long as the head and tail both; how long is the fish?

Note. — The head and tail = the body; the head and tail = the head + the head + $\frac{1}{2}$ the body; then, twice the head = the body — $\frac{1}{2}$ the body = $\frac{1}{2}$ the body; and the length of the body = 4 times the length of the head.

9. Bought a watch, chain and key, for $\$125$; the watch cost 5 times as much as the chain, and the key cost 95 per cent. less than the watch; what was the cost of each?

10. In what time will a given principal double itself, at 5½ per cent. interest?

11. A man can dig a ditch in 3 days, and his son in 5 days; in what time can they dig it together?

12. Jason has 25 per cent. more money than Edward; what per cent. less has Edward than Jason?

13. In how long a time will $60 gain $6.30, at 7 per cent. interest?

14. How much water must be mixed with 10 gallons of brandy, worth $8 a gallon, that the mixture may be worth only $7 a gallon?

15. A and B set out to travel round a certain island which is 20 miles in circuit; A travels at the rate of 5 miles an hour, and B 7 miles an hour; how long will it take B to overtake A?

16. What cost 69 yards of cloth, at 33 cents a yard?

NOTE. 33 is ⅓ of 99, and 99 is 100 less 1.

17. If $100 have been borrowed at 6 per cent. interest, for 1 year, how long must $250 be loaned at the same rate per cent. to requite the favor?

18. At what per cent. interest will $30 become $60 in 12½ years?

19. If you should purchase a lot of sheep, at $2.50 apiece, and should lose ¼ of the number, at what price each must you sell the rest so as neither to gain nor lose?

20. A man left ⅓ of his estate to his wife, ¼ of the remainder to his son, and ⅔ of the remainder to his daughter; the three legacies amounted to $900; how much was the whole estate?

21. A lady having a number of peaches, gave away ¼ of them, and ⅖ of the remainder, and had 27 left; how many had she?

22. Four men rent a field for $16; A puts in 6 cows, B 8 cows, C 4 cows, and D as many cows as his paying $\frac{1}{10}$ of the rent entitles him to; what part of the rent did A, B and C each pay, and how many cows did D put in?

23. Sold a watch for $35, and thereby lost 25 per cent., when there ought to have been gained 30 per cent.; how much was it sold below its proper value?

24. Sold a cart for $30, and thereby lost 20 per cent.; at what price should it have been sold to have gained 20 per cent.?

25. If 4 horses consume 21 bushels of grain in 6 days, how many bushels will 8 horses consume in 12 days?

26. A circular garden is 3 rods in diameter; what is its circumference?

Note. — The circumference of a circle, or the distance round it, is $\frac{22}{7}$, nearly, of its diameter, or the distance across it through the centre.

27. A certain circular pond is 11 miles in circumference; what is its diameter?

28. The sum of two numbers is 19, and twice the first added to 5 times the second is 74; what are the numbers?

29. Williams and Brown enter into business together; Williams puts in 4 times $\frac{1}{2}$ of what Brown does; they both put in $1680, and gain $700; what is each one's share of the gain?

30. A barrel of flour and a cord of wood cost $17, and 2 barrels of flour cost $10 more than a cord of wood; how much does each cost?

31. A hare starts 25 leaps in advance of a hound, and takes 4 leaps to the hound's 3; but 2 of the hound's leaps equal 3 of the hare's; how

many leaps must the hound take to overtake the hare?

32. If a flag-staff 30 feet in height at a certain hour casts a shadow of 20 feet, what must be the height of that staff which at the same time casts a shadow of 25 feet?

33. Two men have a flock of sheep; A has 15 more than half of the number that B has, and both have 54; how many have each?

34. Smith and Robinson go into partnership, each putting in $300; afterwards Smith puts in twice as much, and Robinson 3 times as much as before; what share of the capital has each contributed?

35. The sum of two numbers is 24, and 3 times the first, less twice the second, is 17; what is each?

36. The circumference of a circle is 22 inches; what is its area?

Note. — The area of a circle is equal to half the diameter multiplied by half the circumference.

37. How much greater area has a circle 22 inches in circumference, than a square of the same perimeter?

38. Bought 60 apples at 5 for 2 cents, and sold half of them at 2 for a cent, and half at 3 for a cent; how much was the gain?

39. A man being asked what time it was, answered that the time past noon was $\frac{1}{4}$ of the time past midnight; what time was it?

40. A man being asked the time of day, answered that the time past noon was equal to $\frac{4}{5}$ of the time past midnight; what time was it?

41. A man said that $\frac{1}{2}$ of the time past noon was equal to $\frac{1}{4}$ of the time to midnight; what was the hour?

42. A and B start from the same point, and travel in the same direction around a square, each side of which measures 5 miles; A travels at the rate of 4½ miles an hour, and B at the rate of 3 miles an hour; in what time will they be together again? How many miles will each have travelled? How many times will each have been around the square?

43. There is a cask containing 50 gallons of wine; if one half of the wine be drawn off and an equal quantity of water be added, and then one fifth of this mixture be drawn off and the same quantity of water poured in, how many gallons of wine and how many of water are there in the cask?

44. A cask capable of holding 75 gallons, contains 50 gallons of wine; if enough water be poured in to fill the cask and one third of the mixture be drawn off, and then 10 gallons of water be poured in and one sixth of the mixture drawn off, how many gallons of wine and how many of water remain in the cask?

45. A wolf can eat a sheep in 2½ days; a hound can eat it in 3¼; and a mastiff, in 4 days; after the wolf has eaten ½ of a day and the hound ¼ of a day, how long will it take the hound and mastiff together to eat what remains?

46. A bought a horse for 25 per cent. less than his real value, and sold him to B for 25 per cent. more than his value; how much per cent. did A make on his purchase? How much per cent. would B lose were he to sell the horse for the same price that A gave?

47. A lady has 2 silver cups, and but one cover for both; the cover weighs 10 ounces; now, if the cover be put on the first cup, it will make the weight double that of the second, and if the cover

be put upon the second, it will make the weight triple that of the first; what is the weight of each?

48. There is a bin containing 30 bushels of wheat; if I take from it 10 bushels and add 10 bushels of rye, then take away 10 bushels of the mixture and add 10 bushels more of rye, then take away 10 bushels of this mixture and add 10 more of rye, supposing the wheat and rye to have been thoroughly mixed each time, how many bushels of wheat, and how many of rye would the bin contain?

49. If I should each time add the 10 bushels of rye *before* taking out the 10 bushels of the mixture, and perform the operation three times, as in the last example, how many bushels of each would remain in the bin?

50. If a merchant purchase goods for cash to the amount of $500, when money is worth 2 per cent. a month, what sum will he gain by selling the goods at the end of 4 months, but on 3 months' credit, at an advance of 20 per cent. upon the cost?

51. A father said to his son, "4 years ago I was 3 times as old as you, but 8 years hence I shall be 2 times as old as you;" what was the age of each?

52. A gentleman let $\frac{2}{3}$ of his money at 5 per cent., and the remainder at 6 per cent.; the interest amounted to $90; what were the sums let?

53. After spending $\frac{2}{3}$ of my money, I found if I had spent 1\frac{2}{3}$ less, I should have spent just $\frac{1}{2}$ of my money; how much did I have?

54. A grocer has two kinds of tea, one of which is worth 40 cents a pound, and the other 50 cents a pound; how many pounds of each must be taken to form a chest of 40 pounds, which shall be worth 44 cents a pound?

55. The sum of three numbers is 18, the sum

of the first and second is equal to the third, and half the sum of the first and third is equal to the second; what are the numbers?

56. A farmer employed 3 men and 3 boys one day for $5, and another day, at the same wages, 4 men and 6 boys, for $8; what was the daily wages of each?

57. What number is that to which if 3 and 14 be separately added, the first sum will be ½ of the second?

58. Three men, A, B and C, each, have a sum of money in their pockets; A has $3, A and C together have 3 times as much as B, and B and C together have 11 times as much as A; how much have B and C respectively?.

59. Smith, Jones and Brown, each, have a sum of money at interest at 5 per cent., and these sums are to each other as ½, ¼ and ⅙, respectively; the annual income of the three sums taken together is $90; what is the principal that each one has at interest?

60. Three men hired a pasture for $76; at first A put in 4 horses and 8 cows; B 6 horses and 12 sheep; afterwards, when the grass was half eaten up, C put in 24 sheep; now, supposing every horse eats, in a given time, as much as 4 sheep, and every cow as much as 3 sheep, what ought each man to pay?

61. Robinson, Savage and Harrison, agree together to do a piece of work; they are to receive for it $200, to be divided in the proportion of 5, 4 and 3, respectively, for the same amount of work; but Robinson, whose labor is worth most, is absent ⅓ of the time, and Harrison, whose labor is worth least, is absent ⅓ of the time; how should the money be justly divided among them?

THE END.

GREENLEAF'S
SERIES OF MATHEMATICS.

THE Publishers of this Series, in its present revised and much improved form, would call the attention of Teachers, School Directors, and others interested, to its several important distinctive characteristics:

The ARRANGEMENT *of its several parts and subjects is lucid, progressive, and strictly philosophical.*

The RULES, DEFINITIONS, *and* ILLUSTRATIONS, *are expressed in language, simple, clear, concise, and accurate.*

The PROBLEMS *are of a practical nature, tending to interest the pupil, exercise his ingenuity, and secure useful mental discipline.*

It is a CONSECUTIVE SERIES, *adapted to Primary, Intermediate, Grammar, and High Schools; and is in accordance with the best modern methods of instruction.*

The MECHANICAL EXECUTION *is neat and durable, an important consideration, too often disregarded.*

I. THE NEW PRIMARY ARITHMETIC

Constitutes attractive and interesting *First Lessons* in numbers, and contains all the pupil needs as preparatory to the next book in the series.

In order that the reasoning of some of the principal processes might be the more apparent to the beginner, pictures of objects have been, to some extent, introduced. · After these, counters are employed, as unit marks ; and then follow lessons without any such aids, that the learner may early acquire the habit of depending upon mental resources alone for the solution of problems.

II. THE INTELLECTUAL ARITHMETIC,

As an advanced course of exercises, inductive and analytic, it is thought, will fully meet the requirements of the highest standard of mental culture. It has been the constant aim of the author, in its preparation, to unfold inductively the science of numbers in such a series of progressive intellectual exercises as should awaken latent thought, encourage originality, give activity to invention, and develop the powers of discriminating justly, reasoning exactly, and of applying readily results to practical purposes. The advanced exercises in the fundamental processes of the science, given toward the end of the book, constitute a feature peculiar to this work.

III. THE COMMON SCHOOL ARITHMETIC

Contains *all the important rules of common arithmetic, with their practical applications,* and is ample to prepare the student for all ordinary business transactions. It is a *complete system in itself,* though not so extensive as the National Arithmetic.

IV. THE HIGHER ARITHMETIC, or NATIONAL ARITHMETIC,

Contains a greater amount and variety of matter strictly connected with the science, than will be found in any other treatise of the kind. It embraces a large amount of *mercantile* information, not usually included in works of this nature, but important to be possessed by all who are destined for the *warehouse* or *counting-room.* As a text-book for *advanced* and *normal classes* it has no equal.

V. THE TREATISE ON ALGEBRA

Furnishes what has been hitherto much desired, a thorough practical and theoretical text-book, suited to the wants of Advanced Schools, and Academies, in a single volume of convenient size. Very comprehensive in its plan and details, and progressive in its gradation of problems, it occupies the ground sometimes given to two different books. Its several *demonstrations,* especially those connected with the *roots,* the *method of solving cubic equations by completing the square,* and the very complete *Table of Logarithms* at the end of the volume, are among its useful distinctive features.

The book has now been fully tested in the school-room, and the testimony of teachers is that its merit is fully equal to that of the arithmetics by the same author, and to which it proves the best and most appropriate sequel.

VI. THE ELEMENTS OF GEOMETRY,

WITH PRACTICAL APPLICATIONS TO MENSURATION. Designed for Academies and High Schools. 320 pp. 12mo.

This work has been prepared with great care, and it is believed will fully meet the demands which it is intended to supply in the series.

KEYS TO THE COMMON SCHOOL, NATIONAL ARITHMETIC, AND ALGEBRA, containing Solutions and Explanations, *for Teachers only.*

Two editions of the NATIONAL ARITHMETIC, and also of the COMMON SCHOOL ARITHMETIC, one containing the ANSWERS to the examples, and the other without them, are published. Teachers are requested to state in their orders *which edition* they prefer.

☞ Greenleaf's Arithmetics and Algebra are no untried books, or of doubtful reputation. No other works of the kind have, in the same time, secured so general an introduction, in all parts of the United States, or been as highly commended by eminent teachers and mathematicians.